Further praise for
Growing a Revolution

Soil's greatest living advocate, David Montgomery, has done it again. *Growing a Revolution* proposes the radical idea that by improving soil health, we can heal not only the earth but ourselves as well. A call to action that underscores a common goal: to change the world from the ground up."   —Dan Barber, chef and author of *The Third Plate*

"A wonderful read on how to make soil rich and prosperous!"
                   —Estella B. Leopold, author of *Stories from
                         the Leopold Shack: Sand County Revisited*

"Montgomery has written another classic. *Growing a Revolution* is one of the most important books ever written—an engaging and revealing service to human society and our planet."
                   —Amir Kassam, professor of agriculture, policy
                         and development, University of Reading, UK

"Growing a Revolution presents a clear-eyed examination of a solution to the challenges we face in feeding the world. A joy to read with the bounce and flow of a great biography. I couldn't recommend it more."
                   —Jerry Harrison, keyboardist and guitarist, Talking Heads

"From Plato to FDR, from George Washington to Gabe Brown, Montgomery shows how all roads lead to the soil—and the potential it holds to redress some of our greatest challenges in the twenty-first century."
                   —Woody Tasch, founder of Slow Money and author of
                         *Inquiries into the Nature of Slow Money: Investing
                         as if Food, Farms, and Fertility Mattered*

"Loved the book! Ambitious and thought provoking. A fascinating, thoughtful, and hopeful adventure."

—Diana Wall, past president of the Ecological Society
of America and director of the School of Global
Environmental Sustainability at Colorado State University

"An important masterpiece, *Growing a Revolution* is one of the most practical and prescient descriptions of the food and agriculture transformation we must now accomplish. Should be required reading for anyone interested in the future of food and farming."

—Frederick Kirschenmann, author of *Cultivating an
Ecological Conscience: Essays from a Farmer Philosopher*

"A wise and grounded book—restored soils are the solution."

—Jules N. Pretty, professor of environment
and society, University of Essex, UK

GROWING A REVOLUTION

Also by David R. Montgomery

*The Hidden Half of Nature: The Microbial Roots
of Life and Health* (with Anne Biklé)

The Rocks Don't Lie: A Geologist Investigates Noah's Flood

Dirt: The Erosion of Civilizations

King of Fish: The Thousand-Year Run of Salmon

GROWING
A REVOLUTION

Bringing Our Soil Back to Life

DAVID R. MONTGOMERY

W. W. Norton & Company

Independent Publishers Since 1923

New York · London

For information about permission to reproduce selections from this book,
write to Permissions, W. W. Norton & Company, Inc.,
500 Fifth Avenue, New York, NY 10110

For information about special discounts for bulk purchases, please contact
W. W. Norton Special Sales at specialsales@wwnorton.com or 800-233-4830

Manufacturing by LSC Communications Harrisonburg
Book design by Kristen Bearse
Production manager: Julia Druskin

Library of Congress Cataloging-in-Publication Data

Names: Montgomery, David R., 1961–
Title: Growing a revolution : bringing our soil back to life / David R. Montgomery.
Description: First edition. | New York : W.W. Norton & Company, [2017] |
Includes bibliographical references and index.
Identifiers: LCCN 2016055810 | ISBN 9780393608328 (hardcover)
Subjects: LCSH: Soil ecology. | Soil biology. | Soil restoration.
Classification: LCC QH541.5.S6 M66 2017 | DDC 577.5/7—dc23 LC record
available at https://lccn.loc.gov/2016055810

W. W. Norton & Company, Inc.
500 Fifth Avenue, New York, N.Y. 10110
www.wwnorton.com

W. W. Norton & Company Ltd.
15 Carlisle Street, London W1D 3BS

2 3 4 5 6 7 8 9 0

For the innovative farmers restoring life
to their land so the future can eat.

Upon this handful of soil our survival depends. Husband it and it will grow our food, our fuel, and our shelter and surround us with beauty. Abuse it and the soil will collapse and die, taking humanity with it.

—*Vedas, Sanskrit Scripture, 1500 B.C.*

CONTENTS

INTRODUCTION

There's a revolution brewing—a soil health revolution. Since the dawn of agriculture, society after society faded into memory after degrading their soil. But we need not repeat this history on a global scale. For while the problem of soil degradation remains the least recognized of the pressing crises humanity faces, it is also one of the most solvable. Are you ready for an optimistic book about the environment?

A growing movement of innovative farmers has laid the foundation for this revolution. They are upending conventional ideas and changing practices in ways that leave the soil better off rather than worse off under intensive cultivation. I admit I was skeptical when I first started to look into all this. But what I found convinced me of the transformative potential of adopting a simple set of agricultural practices that together define a new philosophy of farming that merges ancient wisdom and modern science.

This book is the story of my journey to meet these farmers and learn how they build fertile soil as an integral part of how they farm. But there's more to this new breed of regenerative farmer. The secret to their success is that they are also maintaining or increasing their yields *and* increasing their profits. The extra money in their pockets comes from spending less on fossil fuels and agrochemicals. They replace these costly inputs with practices that cultivate diverse communities of soil life that efficiently deliver nutrients, minerals, and other compounds that crops need to grow while fending off pests and pathogens.

The principles underlying the practices of the entrepreneurial and practical farmers leading this revolution can work on farms big, small, high-tech, low-tech, conventional, and organic. And their focus on soil health comes with a silver lining. For changing the way we think about—and treat—soil offers simple, cost-effective ways to help feed the world, cool the planet, and bring life back to the land.

As a geologist I never thought I'd tour the world talking to farmers, let alone start such an expedition in a natural history museum. . . .

I circled behind the elephant, squinted into the overhead lights, and stuck a knife deep into the tasty blue cheese beside the creature's feet. Anne, my wife, was stalking a glass of wine on the far side of the buffet. We were still recovering from our cross-country flight, relaxing in a rotunda of the National Museum of Natural History. I was blissfully unaware that a fertilizer industry lobbyist, of all people, was about to inspire me to write another book—something I'd just sworn to think long and hard about before tackling again.

It was 2008, and I'd received an invitation to speak at a National Research Council–hosted symposium being held in conjunction with *Dig It! The Secrets of Soil*, a new Smithsonian exhibit. The purpose of the symposium was to raise awareness around the issue of soil degradation, a topic close to my geological heart. Plus, I'd been asked to advise them on the exhibit. Naturally, I was curious to see how it turned out.

The most striking display that night was a set of wall-mounted slabs of excavated soils mounted in sturdy wooden frames that covered the walls from floor to ceiling. Fifty side-by-side panels, one from each state in the country, formed a massive patchwork quilt. Arranged in alphabetical order, they cast an earth-toned rainbow—tan from Arizona, mocha from Colorado, black from the Dakotas, and red from Hawaii.

Arranged this way, it was difficult to see a pattern in the colors. I couldn't make geographical sense of it, until I got to the three-panel block of deep-black soil from Illinois, Indiana, and Iowa. Then I rearranged the display in my mind's eye as a map of the United States and it all made geographic sense—the deep black of the plains separating the tan of the Western desert from the rusty red of the Southeast. As much as this colorful wall entranced me, I wondered how many of the other attendees, absorbed in watching videos, pushing buttons, and navigating kid-friendly displays, had noticed the regional character of American soil, obscured as it was by alphabetical arrangement.

Back in the rotunda with the other guests, we continued to enjoy the generous spread of wine and hors d'oeuvres courtesy of the event's sponsor, the Fertilizer Institute. Their speaker began, expounding on the pressing need to use chemical fertilizers. Organic farming could not possibly feed the world, he intoned. The deluded advocates of organic practices offer a recipe for mass starvation. Fertilizers had saved the world by doubling crop yields in the twentieth century. Now they would save us again. As I looked across the room beneath the elephant's belly, the institute's prominently displayed NOURISH, REPLENISH, GROW slogan seemed less innocuous than it had fifteen minutes before.

I had recently become immersed in this very topic, reading papers and studies that reported cases of organic farming matching conventional crop yields. I knew that fertilizers could boost crop yields on degraded, nutrient-poor soils, but they don't help that much on already fertile land. I had started to suspect that the oft-cited conclusion that conventional farms outproduced organic ones depended on the state of the land, as well as the specific practices of those who farmed it. Some studies found that organic practices were as productive as conventional ones, and more profitable because they didn't use expensive chemical inputs—like fertilizers. This research had left me wondering if organic food is generally more expensive because it costs more to grow, or because people are willing to pay more for it. And if the latter were true, then shouldn't an increase in supply relative to demand bring the price down, making it a more viable option for a much larger population if more farmers were to go organic?

Not surprisingly, such ideas were not highlighted in the evening's program. And as the spiel from the sponsor's representative came to a close, I couldn't stop thinking about the gulf between what I'd just heard and what I'd been reading.

The next day at the National Academy of Sciences, a series of experts also painted a different picture, stressing the importance of using organic matter to maintain soil fertility. World-renowned scientists talked about how soil conservation and soil health are vital for feeding a growing population over the long run, especially after we burn through the cheap, abundant fossil fuels that have enabled reliance on industrially produced fertilizers.

Ohio State University soil scientist Rattan Lal presented an idea that particularly intrigued me: put carbon back into the ground to both remove it from the atmosphere and to improve soil fertility. A soft-spoken gentleman, he didn't look particularly revolutionary in his dark suit and tie as he stressed the urgency of sustaining soil quality on a warming planet. But his message was as radical as his manner professorial. He argued that conventional farming practices degraded carbon-rich soil organic matter, not only reducing fertility but contributing to global carbon dioxide (CO_2) emissions. Putting more carbon back into the ground through increasing the amount of organic matter in agricultural soils would both enhance soil fertility, and thus food production, and could go a long way toward offsetting CO_2 emissions.

Of course, there was a catch. Doing so would require wholesale changes to our agricultural practices. I began to wonder whether the organic-versus-conventional distinction was too simplistic. Maybe mineral fertilizers are like many other tools in that it's the way they are used that governs whether they enhance or degrade the soil. Could building soil health depend more on adopting practices that protect the soil from erosion and build up soil organic matter than on forgoing agrochemicals? As I thought about how to do this on a global scale, I had no idea that I'd already begun following in Lal's footsteps and would soon embark on a journey that would take me to farms around the world. By the end, I'd find cause for optimism. For it turns out that we *can* change the practice of agriculture in ways that leave soil better off instead of degrading it. And doing so would help us overcome the daunting challenges of feeding the world and cooling the planet.

GROWING A REVOLUTION

1

FERTILE RUINS

Civilization itself rests upon the soil.

—Thomas Jefferson

What if I told you there was a relatively simple, cost-effective way to help feed the world, reduce pollution, pull carbon from the atmosphere, protect biodiversity, and make farmers more money? If this was true, you might assume that governments around the world would race to embrace it. Well there is, and they aren't. Not yet anyway.

Why not? Because the solution challenges a century of conventional wisdom and powerful commercial interests, and requires a profound shift in how we think about and treat the least glamorous resource of all—the soil beneath our feet.

But before we get to the good news, let's look at where we are and how we got here. It's not a rosy picture. We have already degraded at least a third of the world's agricultural land. A *third*. And though we rarely hear about it, degradation of farmland presents as great a threat to civilization as global conflict, our exploding population, climate change, and dwindling supplies of fresh water.

In 2015, the United Nations Food and Agricultural Organization released a report by a consortium of scientists from around the world that estimated soil degradation erodes almost half a percent of our global crop production capacity each year. By any accounting, such a trend can't go on for long without serious consequences. Indeed, we are already seeing foreign interests buy up farmland in developing nations. Not to grow food for the local populace, but for export back home. With food shortages already fueling violence in drought- and conflict-plagued regions like Nigeria, Somalia, and Syria, such arrangements will not favor global stability.

Among the classical elements of earth (soil), air (climate), fire (energy), and water, it is the first that consistently gets overlooked or short-changed in public discourse and policy. Yet we might consider fertile soil as the ultimate strategic resource. For there is no substitute as there is for oil, and it cannot be distilled, as fresh water can be from seawater, nor cleaned by filters as air can. And because we do not fully recognize the value of what's beneath our feet, we risk repeating ancient mistakes on a global scale.

From the Roman Empire to the Maya and Polynesia's Easter Island, one great civilization after another sank into poverty and eventual demise after destroying their topsoil. But the effect of soil degradation on human societies is not simply a historical footnote. We too are facing challenges these once-thriving societies faced, only now simultaneously from North Carolina to Costa Rica, India, and Africa. And if we don't implement solutions soon, we'll find ourselves in the same dire circumstances globally as did our regional predecessors. For with even less fertile soil, how will we feed several billion *more* people in the future?

Unlike other environmental problems such as dwindling water supplies and loss of forests, the degradation of soil fertility has gone relatively unnoticed. It happens so slowly that it rarely becomes the crisis du jour. Therein lies the problem. The once-Edenic, now-impoverished places that spawned Western civilization illustrate one of history's most underappreciated lessons: societies that don't take care of their soil do not last. We cannot afford to repeat the mistakes of the past now that there's nowhere left to go. We've already farmed, developed, or degraded and abandoned pretty much all the land well suited for agriculture over the long run.

Today, however, feeding the world is more of an economic and distribution (and therefore political) problem than an agronomic one.
Even with the loss of a substantial amount of fertile land, we currently
harvest enough food to feed everyone, in principle if not in practice.
But we can't count on being able to meet the demand for long if fertile
cropland continues to shrink while the global population continues
to rise.

Of course, there are other dimensions to the problem of global hunger. Aside from population growth, there are issues of land tenure and
how much of our harvests goes to feed livestock and cars (biofuels).
But a far-too-neglected factor in thinking about how to feed the world
of tomorrow is the potential to restore land to agricultural productivity. Could we really restore fertility to degraded farmland? If so, how
much—and how fast?

A growing movement of farmers around the world is starting to
turn ancient patterns around, restoring life and health to their land.
Yet we don't hear much about this movement. With no products to
sell, they tell a different story than that of conventional interests. This
movement is gaining momentum because farmers who adopt its practices save time and money, and many increase harvests. These practices
can work on large and small farms across the technological spectrum,
from enormous farms in the Dakotas to small subsistence farms in
Africa. And, if implemented well and globally, they could solve one of
civilization's oldest problems.

THE PROBLEM OF THE PLOW

I confess I never thought I'd write an optimistic book about the environment. For many years I was a dark green ecopessimist convinced
humanity was rushing headlong into self-inflicted disaster. While I
still harbor some such fears, I've become far more positive about our
long-term prospects. Over the past few years, I've traveled extensively,

meeting visionary farmers who are restoring life and fertility to their land. These experiences convinced me that it's possible not only to restore soil on a global scale, but to do so remarkably fast.

At least I hope it is, since we face the confluence of the end of cheap oil, continued population growth, and a changing climate over the coming century. How farming will adapt remains uncertain, as political, economic, and environmental interests push competing visions, policies, and agendas. No matter how all this plays out, it will shape the fate of nations and define the world we leave for generations to come.

My perspective on this issue started to change a decade ago, after I did something some colleagues might consider unpardonable—I wrote a book about soil and titled it *Dirt*. You see, soil scientists consider it blasphemous to call soil *dirt*. This is because there are very important differences between soil and dirt. For one, soil is full of life, dirt is not. So why would a geologist like me write an irreverently titled book about the importance of what covers up rocks? While my primary focus of study is how landscapes are shaped by natural processes and changed by people, over the course of examining the evolution of landscapes around the world, I came to see how soil erosion and degradation influenced human societies.

Some geologists argue that people, directly and indirectly, now move more earth around than nature herself. Earth scientists have even proposed a new epoch, the Anthropocene, or "Age of People." Although we argue about when this epoch started, it is perfectly clear that of all our world-changing inventions, the plow was, and remains, one of the most destructive.

Yes, you read that correctly. The *plow*. That iconic symbol of our agricultural roots that helped launch civilization as we know it. The plow enabled few to feed many and set the table for the rise of commerce, city-states, and hierarchical societies with priests, princes, politicians, and all the rest of us who don't farm. The problem, in a nutshell, is that the plow makes land vulnerable to erosion by wind and rain.

Consider that you rarely see much bare earth in natural grasslands or forests. Where she can, nature clothes herself in plants because ground stripped of plant cover—like a just-tilled field—erodes faster than soil forms. Tillage also pushes soil downhill with each pass of the plow. Thus, when plowed generation after generation, hillslopes gradually—and sometimes not so gradually—lose their natural endowment of topsoil. And so, storm by storm, with each pass of a plow, a millimeter at a time, the land slowly loses fertility.

Studying how erosion shapes topography in different settings around the world, I noticed how the prosperity of societies can mirror the state of their land. The point was driven home for me when I was doing fieldwork in the Amazon. Driving through an area of rainforest freshly cleared for subsistence farming, I saw how fast bare fields eroded down to nutrient-poor weathered rock. Where this happened, impoverished families could barely scratch a living from the earth. The farmers would soon move on to clear fresh fields from the rainforest, as ranchers moved their cows in behind to graze abandoned fields. This became a cycle of destruction with no end in sight. On another trip, this time to the South Pacific island of Mangaia, I saw how badly eroded soils barely fed a small population with a long history of fighting over ever-fewer productive fields.

Through fieldwork spanning three decades and six continents, I realized that long-cultivated regions that had lost their topsoil remained impoverished as a result. Telltale signs are etched in ragged gullies and slopes with subsoil exposed at the surface. The poor fertility of the soil that remains on the land is harder to see.

However, it's worth noticing—and reversing. For restoring the soil can help address the fundamental challenges of water, energy, and climate, as well as a number of important environmental and public health problems. Nitrogen pollution, born of our dependence on fertilizers, is affecting urban water supplies in the Midwest and creating a great dead zone in the Gulf of Mexico off the mouth of the Mississippi River. Algal blooms from excess phosphorus in agricultural runoff kill fish in the Great Lakes. Direct exposure to insecticides and indirect effects of herbicides that kill their food source contribute to crashing populations of pollinators, like bees and Monarch butterflies, with dire implications for crop production and biodiversity. Wholesale reliance on agrochemicals directly affects human health too, as increased risk of depression and certain cancers are associated with pesticide exposure. Restoring healthy, fertile soil would cast a broad net, helping to address all these problems. So how feasible is it?

After writing *Dirt*, I received invitations to speak about the history of soil loss and degradation at more farming conferences than I can remember. This gave me opportunities to travel to places I wouldn't otherwise go (geologists usually gravitate toward mountains rather than to flat farmland), and the chance to meet innovative farmers I wouldn't normally encounter. At first, I didn't fully appreciate this opportunity. But after hearing one story after the next of how farmers revived degraded land, I started seeking out their opinions on this pressing issue. In doing so, I began to realize that I shared more common ground with farmers than I thought. Many of them saw the destructive effects of plowing as clearly as I did, if not more so.

In 2010, Guy Swanson invited me to speak at a farming conference in Colby, Kansas. His company sells an attachment to no-till planters that helps farmers reduce the amount of fertilizer they use. No-till farmers don't plow, they use specialized planters that open a narrow slot in the soil about the width of a kernel of corn. Seeds drop down into the slot, disturbing much less of the surrounding soil than plowing it up would.

Swanson's system injects a uniform amount of fertilizer adjacent to and below each just-planted seed, putting nutrients right where plants need them—and only there. This uses far less fertilizer than spraying it all over the field. The farmer saves money and fewer chemicals run off to pollute streams, lakes, and oceans. That sounds like a win-win, except of course to fertilizer companies. Swanson had seen me talk at a no-till farming conference and wanted me to come speak about the civilization-killing problem of soil erosion to potential customers contemplating a shift to no-till methods and precision fertilizer use.

As Swanson and I drove into Colby in his well-traveled white Impala, an enormous billboard welcomed us with a larger-than-life, hippy-looking Jesus peering out over a field of wheat. I started to worry about how Swanson's audience would receive a professor from the far left corner of the country talking about how the plow had ruined the land for society after society throughout history. Would I be run out of town before or after lunch?

As I ended my talk, I looked out on a sea of baseball hats. One elderly fellow in the middle stood up, stuffed his hands down into his pockets, and said he'd taken one look at me and didn't think I could possibly say anything worth listening to. I braced myself for what was to come. But then he surprised me. He said the more I'd talked, the more sense I'd made. He'd seen what I was talking about on his farm. It no longer had the rich fertile topsoil his grandfather had plowed. Something needed to change if his own grandchildren were going to prosper working his land.

Time and again, at one farming conference after another, instead of walking out or lobbing verbal grenades at me, farmers readily acknowledged the possibility that plowing resulted in long-term damage to the soil. A surprising number said they knew this to be true from firsthand experience. Older farmers would share stories about how their soil quality had gone downhill over their lifetimes, too slowly to notice year to year, but plain as day in retrospect. One after another piped up to say that they'd noticed their soil decline under the now-conventional marriage of the plow and intensive fertilizer and agrochemical use.

In hindsight, I really shouldn't have been surprised that farmers recognized the twin problems of soil loss and degradation. After all, who knows the land better than those who work it for a living?

But the trip with Swanson was the first time I'd really had a chance to talk with farmers who were enthusiastic about reducing chemical inputs to save—and thereby make—money. Most liked the potential of using technology and precision fertilization to do more with less. And, while many expressed a desire to stop soil erosion, they all loved the idea of lowering next year's fertilizer bill without lowering crop yields. So I started paying more attention to what individual farmers thought it would take to carry on farming well into the future. I asked them what they were doing—and how they were doing it. It didn't take long to see common threads running through their answers.

When Swanson and I left Colby and drove south toward the heart of the 1930s-era Dust Bowl, I noticed both prosperous and dilapidated John Deere dealerships, depending on what part of Kansas we were in. Some of those we sped past held rows of shiny new green machines in clean, well-kept yards. Others featured broken-down fences surrounding lots full of rusted equipment. When I asked Swanson about this striking disparity, he said the thriving dealerships were in counties where farmers were going no-till.

This filled me with hope. After all, I figured the only way we could shift the dominant paradigm, as professors like to say, is to entice farmers with a combination of good stewardship and economic gain. If farmers saw that protecting and improving their soil left more money in their wallets, then we might break the age-old cycle of soil degradation that has destroyed societies throughout history.

I began to wonder what it would actually take to generate a resilient, productive, and permanent agriculture. I doubted there was a simple one-size-fits-all-farms answer. And I knew the answer wasn't simply organic farming. Many, if not most, organic farmers plow to suppress weeds and prepare the ground. I realized that the basic question that society needs to focus on is how farmers of all stripes can forgo the plow and leave their soil better off after a crop is planted and harvested—over and over again.

And one way or another, change will come—and already has. Think about where you grew up and the amount of land developed since you were a kid. Orchards or fields, rolling or flat, we've all seen productive farmland ripped up to make way for suburbs and strip malls. Back in the 1980s, as a newly minted geologist right out of college, I worked as a foundation inspector for a geotechnical engineering firm in the Bay Area. I quickly learned that carting off the topsoil was the first thing contractors did as they turned then-rich farmland into today's Silicon Valley. Loose, fertile soil doesn't offer much support for foundations. It settles if you park a building on it. So, if you want to build a business park, the topsoil has to go. On project after project, I watched rich, black earth get loaded onto trucks and hauled off to be used as fill. None of those acres will grow food for generations to come. And there will be a lot more of us before our global population tops out sometime later this century.

Learning the story of past societies and seeing wholesale destruction of farmland made me seriously doubt humanity's ability to feed the future. After all, it takes nature centuries to build an inch of fertile top-soil, and we were inadvertently on track to destroy it all in a few generations. This sure seemed like a story that wouldn't end well. Until, that is, my wife decided we needed a garden.

Right after I received tenure, Anne and I bought a house in north Seattle. The yard was a ratty, century-old lawn with nary a worm residing in the lifeless dirt below. But there was space for the garden Anne dreamed of and we weathered a yard demolition project to start over with a blank slate. And then I patiently endured Anne hauling all kinds of organic matter home to make mulch and compost to improve the soil in our new planting beds. It was only in retrospect that I realized our garden beds were miniature versions of farm fields. After a few years we started to notice the results. As the soil changed color from khaki to chocolate, life seemed to spring from the earth—worms, millipedes, spiders, and beetles. Pollinating insects and birds soon followed. A roar of life emerged from beneath our feet and rippled above-ground, transforming our yard and our view of the world.

We had rediscovered the tried-and-true practice of using organic matter to build fertile soil, leading us down the nearly forgotten path of working *with* nature. Much to my surprise, the soil degradation that had taken down ancient societies was running in reverse right in my own backyard. And it was happening far more quickly than I thought possible. Seeing the transformation of our urban soil firsthand confirmed for me that soil biology is not only central to building soil fertility, but that we can use it to restore soil much *faster* than nature makes it. This was not what I had learned in college, where I'd been taught that soil chemistry and physics governed soil fertility and that nature made soil at a glacial pace.

Long before Anne and I stumbled onto restoring life to our soil, the Dutch began diking off land from the sea and systematically proceeded to build up beach sand into the famously fertile, rich, black soils that still comprise some of Europe's best agricultural land. Their secret? Returning organic matter to the fields.

And long before this Dutch innovation, Amazonian Indians made fertile, black terra preta soil, burying organic waste with charcoal from cooking fires to create patches of rich topsoil around their villages in an environment with naturally infertile soil. In the Andes, the Inca built terraced fields, which after thousands of years of farming now host soils more fertile than the native hillsides. The Incan trick sounded familiar—hauling organic matter back to their planting beds. Across Asia, the long-standing practice of returning manure and "night soil" (human excrement) to the fields followed the same principle. These societies all built and sustained fertile soil through soil-building agricultural practices.

My own experience and looking at such history told me that people can build up soil organic matter, and thus soil fertility, far faster than nature makes fertile soil. But our agricultural policies hold us back, discouraging farmers from using tools right beneath their feet. Yet, I know farming practices can change. After all, they have many times in the past.

A NEW REVOLUTION

A look back at our agricultural past reveals a long series of innovations, and a few bona fide revolutions, that greatly reduced the amount of land it takes to feed a person. These changes led to a dramatic increase in how many people the land could support and a corresponding decrease in the proportion of people who farm. By my reckoning, we've already experienced four major revolutions in agriculture, albeit at different times in different regions.

The first was the initial idea of cultivation and the subsequent introduction of the plow and animal labor. This allowed sedentary villages to coalesce and grow into city-states and eventually sprawling empires. The second began at different points in history around the world, as farmers adopted soil husbandry to improve their land. Chiefly, this meant rotating crops, intercropping with legumes (plants that add nitrogen to soil), and adding manure to retain or enhance soil fertility. In Europe, this helped fuel changes in land tenure that pushed peasants into cities just in time to provide a ready supply of cheap urban labor to fuel the Industrial Revolution.

Agriculture's third revolution—mechanization and industrialization—upended such practices and ushered in dependence on cheap fossil fuels and fertilizer-intensive methods. Chemical fertilizers replaced organic-matter-rich mineral soil as the foundation of fertility. Although this increased crop yields from already degraded fields, it took more money and required more capital to farm. This, in turn, promoted the growth of larger farms and accelerated the exodus of families from rural to urban areas. The fourth revolution saw the technological advances behind what came to be known as the Green Revolution and biotechnology breakthroughs that boosted yields and consolidated corporate control of the food system through proprietary seeds, agrochemical products, and commodity crop distribution—the foundation of conventional agriculture today.

What will the future hold as we burn through the supply of cheap oil and our population continues to rise alongside ongoing soil loss and climate change? A recent study authored by hundreds of scientists from around the world concluded that modern agricultural practices must change once again if society is to avoid calamitous food shortages later this century. Just how worried should we be? Well, consider the fate of Mesopotamia, ancient Greece, or other once-great civilizations undone by their failing land. This time we need to ask what agriculture would look like if we relied on building fertile soil instead of depending on chemical substitutes. What would this new, fifth agricultural revolution look like?

Those at the vanguard invoke a variety of names—agroecology, conservation agriculture, regenerative agriculture, and the Brown Revolution. While proponents of these approaches include those who passionately disagree about the roles of organic practices and genetic engineering in the future of agriculture, I am more struck by the common ground they share in placing soil health at the heart of their practices.[1]

When the United Nations declared 2015 the International Year of Soils, I received more invitations to speak at soil-themed conferences. Interest in soil health had grown rapidly among farmers compared to what I had seen just a few years before. The seeds of optimism took root as I listened to farmers tell of how they changed the way they farmed, restoring life and fertility to their land. After a while, I started to think we might actually get it right this time. Maybe we could reverse the ancient pattern of farming ourselves out of business. Was this the start of a new movement that could rewrite the ending to an age-old tale—one farm at a time?

Seeking to understand what an agricultural revolution centered on soil health might look like, I set off on a trip across several continents to visit farmers who were restoring life to their land. What I learned shattered central myths of modern agriculture and pointed to simple, effective ways to help solve some of our most vexing problems. Individually, their examples illustrate how changes in farming practices can transform both conventional and organic agriculture into more sustainable enterprises. Collectively, they build a compelling case that we can feed the world, start cooling the planet, and restore life to the land.

The story started to come together as farmers from different regions told me their reasons for restoring their soil. Stopping erosion was usually on the list. Saving money by consuming less water, oil, fertilizers, or pesticides always was. Alignment of these objectives offers an opportunity to break free from the cycle of land degradation that doomed ancient societies.

Not all the farmers I met did things the same way. How could they? They grew different crops in different regions with different soil and different climates. Some integrated livestock into their operations. Others favored cover crops. A few, perched in the cabs of space-age prairie crawlers, worked fields stretching to the horizon. Others labored by hand in the tropics to coax sustenance from small plots to feed a single family. As varied as their situations and practices were, they all viewed farming as working *with*, rather than against, nature. When I realized that they all operated according to a common set of principles, I knew that the foundation for a new agricultural revolution had already been laid.

This journey challenged my view of the usual debates—organic farming versus GMOs, conventional farmers versus environmentalists, cows versus the planet. I came away with a new appreciation for how some simple changes in practices can help farmers of all stripes, conventional and organic alike.

Perhaps most striking to me was the way this movement is growing bottom-up, fueled by individual farmers rather than governments, universities, or environmental advocacy groups. Naturally enough, farmers share their experiences of what works well for them, or not. And curious neighbors notice when the fool next door tries something different—and then outproduces them, making more money several years in a row. Eventually, I came to realize that this is precisely why this time around we might actually break the age-old cycle of land exhaustion and societal collapse. It is starting to make economic sense for individual farmers to adopt regenerative practices that reinvest in their primary long-term asset: the fertility of their land.

The singular message that came through loud and clear from farmers I visited was that restoring the productive capacity of the soil could be done quickly and profitably. But it meant doing things differently, a willingness to walk away from conventional practices and to take a chance on the idea that building healthy soil was the best investment a farmer could make. Most of all, it seemed, it took the courage to try new things in the face of regulatory disincentives and skeptical corporate and academic crop advisors. These farmers were not being encouraged to change. They were deciding for themselves that they needed to practice a radically new form of agriculture.

Seeing farmers across a wide range of social, environmental, and economic settings heal and improve their land faster than I thought possible motivated me to write this book. On this journey, I learned that it's not a matter of heeding the sirens of agribusiness versus the prophets of sustainability, or choosing between modern technology and a return to preindustrial farming. That would be far too simple—and misleading. For the most promising way forward lies in the marriage of agrotechnology and agroecology to rebuild soil fertility. Combining ancient insights and modern science, I believe, is the way to both sustain agriculture and use it to help slow climate change.

Most striking to me is how individuals who see the power—and promise—of thinking differently are driving the new movement. To be sure, ongoing technical advances in genetic engineering, precision agriculture, and microbial ecology offer distinct choices in approach, each with its own merits and pitfalls. But I've come away believing that the foundation for the next agricultural revolution will be rooted in how we think about soil. For that colors everything else—especially how we use the knowledge and technology at our disposal.

The degradation of soils and the loss of organic matter are the most underappreciated environmental crises humanity now faces. But the stage is set for ground-up transformational change, as the short-term interests of farmers increasingly align with preserving long-term soil fertility.

It is no coincidence that some of the oldest known works of art are representations of long-forgotten fertility deities. For millennia, people believed that ensuring bountiful harvests depended on humoring fickle gods. But since the days of ancient Egypt, Greece, and Rome, our views of soil fertility have evolved through dramatic shifts in perspective. Now they must do so again if we are to avoid eroding civilization's agricultural foundation.

Though already underway, the revolution still has a long way to go. Like all revolutions, it faces entrenched opposition from powerful interests and conventional thinking. Yet if it succeeds, it could solve one of humanity's most pressing problems: how to keep feeding us all on this lonely rock in space.

2

MYTHS OF MODERN AGRICULTURE

Nothing is so painful to the human mind as a great and sudden change.

—Mary Shelley

Projections of global oil supplies show we either have already passed, or will soon cross over, the peak in crude oil production. As we slide our way down the global supply curve, it will make less and less sense to continue relying on oil. And while there may be vast reserves of shale oil, tar sands, and coal yet to be mined, prominently published projections estimate that even burning a quarter of the world's currently known fossil fuel reserves could alter the world's climate enough to catastrophically reduce crop yields and destabilize agriculture. A 2015 paper published in the journal *Nature* estimated that in order to keep climate change below a highly disruptive 2° Celsius rise, we would need to refrain from using a third of global oil reserves, half of natural gas reserves, and more than 80 percent of coal reserves.

It's no secret that the debate over just how much shale oil, tar sands, and coal we need to leave in the ground to forgo disaster is a controversial one. While agriculture accounts for less than a quarter of global fossil fuel emissions, one way or another farming will be affected. And experimental studies of crop growth predict that the yields of major grains will drop 10 percent for each degree of global temperature rise.

We've already seen what this looks like. The European heat wave of 2003 reduced crop yields by up to 36 percent as the summer average temperature rose 3.6°C above the long-term average for the prior century. So, if we keep burning fossil fuels and adding CO_2 to the atmosphere, it's likely that harvests will decline as population grows. Yet, were we to stop using fossil fuels to make fertilizers and operate farm machinery, we would have to give up what we now depend on to prop up harvests. This dilemma presents a recipe for regional instability and conflict, if not global disaster.

Over the past century, technology reshaped agriculture to the point where modern farming cannot persist without the one thing we're sure not to have in the not-so-distant future: abundant cheap oil. Any way you look at it, if we are to solve both climate change and food-shortage issues, we must wean agriculture off fossil-fuel–intensive methods.

The essential question is not whether agriculture will change, but *how*—and how resilient the land, and thus society, will remain after it does. In an era when we routinely have less than a year's supply of food on the planet, regional crop failures influence global food supplies. This means that agricultural resilience is paramount for maintaining food security and social stability.

The threat of food riots is quite real. Most cities today have a modest supply of food on hand at any time. And history shows that food is the currency of last resort, worth more than cash when both are in short supply. Riots swept ancient Rome when the government failed to provide free bread to the citizenry. Food riots in Paris ignited the French Revolution. But such crises are not just a thing of the past. In 2008, food riots broke out in Asia, Africa, and the Middle East. In 2013, hungry people resorted to violence in drought-ravaged areas of Egypt, Turkey, and Syria; likewise in Venezuela and Somalia in 2015. Starving people do not respect fences or borders.

It is not hard to imagine regional conflicts over fertile land if humanity fails to meet the coming challenge of reliably feeding 10 billion of us. Already many of today's global flashpoints reflect legacies of historical land degradation. For instance, conflict over fertile land has shaped tribal animosities in the Middle East since biblical times. As global population keeps rising, regional food shortages appear inevitable if we continue losing fertile soil.

The Reverend Thomas Malthus, a scholarly nineteenth-century English cleric, infamously predicted that an exponentially mushrooming population would eventually overwhelm steady growth in crop yields. While we might yet prove him right, so far things have not worked out according to the king of pessimists' playbook. For over the course of history, a series of agricultural revolutions allowed our numbers to keep climbing, despite periodic famines and disasters.

Can we pull another rabbit out of our collective hat? Maybe. But isn't it foolish to count on not-yet-existing technologies to solve looming problems? Especially when instead of relying on new technology, this time the key might be a fundamental change in thinking about soil and its fertility.

MODERN MYTHS

The amount of land it takes to feed a person has dropped throughout history, to the point where global agriculture now requires less than half an acre per person. Back in our hunting and gathering days, it took about 250 acres of land to support a single individual. If we still needed that much land per capita, it would take five hundred Earths to feed the current global population. Going back to a preagricultural food supply is simply not an option.

Of course, farming practices that boost crop output in the short term but degrade land over the long haul cannot be sustained either. After all, as Will Rogers famously quipped about land, "They ain't making any more of it." If we are to maintain a global civilization capable of feeding billions of us in perpetuity, we must develop lasting methods of intensive agriculture. The question is, how?

Listening to farmers who are already restoring life to their land unveiled some major myths of modern agriculture for me. They repeatedly told me that they had increased the fertility of their soil and restored profitability to their farms by walking away from conventional practices. Their stories helped shatter the ideas that industrialized agrochemical agriculture feeds most of us in the world today, is more efficient, and is the only way to feed the world of tomorrow. None of these pillars of conventional wisdom are true. So let's look at each a little more closely.

MYTH No. 1: Industrialized Agrochemical Agriculture Feeds the World Today

According to the U.N. Food and Agriculture Organization, family farms produce 80 percent of the world's food, and almost three-quarters (72 percent) of all farms worldwide are smaller than one hectare (about 2.5 acres or the size of a typical city block). In other words, much of humanity eats food grown on small farms. Of course, large industrialized farms allow relatively few people to feed the rest of us living in the developed world. Today, only about 1 percent of Americans are farmers. But most of the world's farmers work the land to feed themselves and their families. So, while industrialized agriculture feeds the *developed* world, it does not feed humanity. Still, we need large-scale agriculture unless we all want to live on—and work—our own farms. But bigger farms are not necessarily more efficient, which leads us to the second myth.

MYTH No. 2: Industrialized Agrochemical Agriculture Is More Efficient

Most industries have economies of scale that lower production costs per unit output for larger-volume operations. But efficiency can also be viewed in terms of input use per unit of production. An authoritative 1989 National Research Council study concluded that "well-managed alternative farming systems nearly always use less synthetic chemical pesticides, fertilizers, and antibiotics per unit of production than conventional farms."[2]

And do larger farms produce more food per acre? No. They may produce particular crops more cheaply, but they don't produce more food overall.

When then U.S. Secretary of Agriculture Earl Butz advised farmers to "get big or get out" in the 1970s, it had more to do with the capital requirements of modern commodity farming than how much food could be produced per acre. Unlike most production processes, farming has an inverse economy of scale in terms of total output. The common misconception that big, mechanized farms produce more food is based on yields per hectare for *individual* crops. Farms that grow a *diversity* of crops produce more food per hectare overall. Small, diversified farms produce more food from a given amount of land than large industrialized monocultures.

When it comes to farming, we've known for decades that bigger does not mean more efficient in this regard. This is no secret, and it's not counterculture propaganda. According to a 1992 U.S. agricultural census report, small farms produce more than twice as much food per acre than large farms do. Even the World Bank endorses small farms as the way to increase agricultural output in developing nations where food security remains a pressing issue.

Another way to think about the efficiency of modern agriculture is that we burn ten calories of fossil fuel to grow one edible calorie. Because of this, it's been said that we are eating oil. But it would be more accurate to say that we are eating natural gas. For industrial fertilizer production not only depends on the ready availability of cheap energy, it also consumes a lot of natural gas as feedstock. It is axiomatic that for any organism to be viable over the long run, it must get more energy from eating than it expends acquiring food. That modern societies don't hold to this simple test of biotic viability should concern anyone with an interest in the future. This leads us to a third myth of modern agriculture.

MYTH No. 3: Intensive Agrochemical Use Will Be Necessary to Feed the World Of Tomorrow

In recent years, the rate of increase in crop yields that characterized the second half of the twentieth century has started to stall. Simply using more fertilizer isn't going to produce bigger harvests since we already add a lot more than our crops take up. And plant response to fertilizers is greatest on degraded soils with little soil organic matter. In other words, adding fertilizers to already fertile soils does not really boost crop yields. The often-repeated line that intensive agrochemical use will be necessary to feed the world overlooks the potential to increase crop yields through adopting practices to restore soil fertility and boost yields on both low-input conventional and organic farms.

Beyond the three foregoing myths, there are several other things worth considering. Organic agriculture can prove as unsustainable as conventional farming when tillage is a regular practice. After all, it was farming without synthetic chemical inputs that impoverished ancient societies. And could there really be enough organic manure available to replace all the chemical fertilizers farmers apply to fields today? How else might we do it? As we'll see, a viable option is to grow cover crops, especially nitrogen-fixing species, as green manure.

In addition, the projected need to dramatically increase global crop production is based on the assumption that Western-style diets rich in grain-fed meat and processed foods will become more prevalent as incomes rise around the world. However, research findings and popular media increasingly link the modern rise in chronic diseases to the Western diet. This has begun to change consumer behavior in the Western world and it is plausible that the people of other nations will not prove as eager to adopt a Western diet as predicted. If both the developed and developing worlds adopted healthier diets with modest meat consumption and an abundance of fiber-rich whole plant foods, it could go a long way toward solving the problem of feeding us all and improving global health.

Another way to slow projected increases in global food needs is to reduce food waste. Thirty to 40 percent of all crops are lost to pests and disease before harvest—despite heavy use of pesticides.[3] And about a quarter of all food produced worldwide gets lost after harvest or wasted between production and consumption. Add those together and half the crops we plant don't end up feeding anyone. The United States alone wastes 133 billion pounds of food each year, more than enough to feed the 50 million Americans who regularly face hunger.

What all this means is that the simple metric of crop yields is too narrow a lens through which to assess strategies for growing enough food for our rising population. So what other factors must we take into account in order to feed the world of tomorrow? Naturally, good places to begin include adopting a healthier diet and reducing waste. Still, the bottom line remains finding and practicing ways to grow more food with fewer inputs. We need to lower the environmental impacts and carbon footprint of agriculture without compromising harvests— and continue doing so in perpetuity. Shifting to low-carbon energy, like wind and solar, is feasible, both technically and economically, if not yet politically. While we may well need to use some fertilizers to feed the world, we will need to figure out how to get by with less of them. And I'm confident we can. I've met farmers who already are doing it.

In fact, the more I learned, the more perplexed I became as to why agricultural policymakers seem to focus exclusively on agrochemistry and biotechnology as the future of agriculture. This is not to say we shouldn't take advantage of technological progress. But we should not let faith in it blind us to effective practices. For restoring life and fertility to the soil is possible right now using existing technology and practices that we know already work to improve soil health. Unfortunately, these practices buck conventional positions that dominate the world of agricultural policy.

THE GMO SIDESHOW

Decades ago, proponents of genetically modified (GM) crops promised increased yields and reduced need for fertilizers and pesticides. Did they deliver?

A 2016 report from the National Research Council Committee on Genetically Engineered Crops found that "nation-wide data on maize, cotton, or soybean in the United States do not show a significant signature of genetic-engineering technology on the rate of yield increase."[4] It's a mixed record for pesticide use. Herbicide-resistant crops led to greatly expanded use of the broad-spectrum herbicide glyphosate since 1996. And the resulting spread of glyphosate-resistant weeds is now leading to increased use of other herbicides as well. In contrast, the introduction of genetically modified Bt crops reduced insecticide use by more than 25 percent, and helped remove from widespread use some of the most environmentally damaging insecticides.[5] Still, overall pesticide use in the United States increased by about 7 percent as a result of adopting GM crops, according to a 2012 study. So claims of increased harvests and reduced pesticide use do not appear to hold up.

Studies also report, however, that GM crops have had some environmental benefits, notably reduced soil erosion from the increase in no-till farming due to simple, flexible, and effective weed suppression using glyphosate. And adoption of BT crops lowered use of some highly toxic, broad-spectrum insecticides. Yet GM crops have also created new and unexpected problems. Just several decades after widespread adoption of GM corn and soybeans, herbicide-resistant weeds and voracious Bt-resistant nematodes (microscopic worm-like creatures that can devour plant roots) are fast becoming serious problems. This means that farmers today are spending more and working harder simply to stay even, fighting battles they can't hope to win and can't afford to lose.

Yet debate over the future of agriculture is misrepresented when cast as the simple choice between organic methods and agrotech approaches like GMOs. Instead, it really comes down to the philosophical rift between agricultural practices based on enhancing nutrient cycling and soil health versus those that mine soil fertility and attempt to replace or compensate for degraded soil health with technology and commercial products. All too often the latter makes short-term economic sense, especially once the soil is already degraded. Therein lies the real dilemma at the heart of our modern crisis: the immediate financial interests of corporations that supply farmers do not necessarily align with our collective interest in maintaining the health and fertility of our agricultural soils.

How might we align farmers' short-term interests with society's long-term needs? By updating traditional ideas into modern practices that define a new system of agriculture. And to do this we need to revisit another myth about soil—that soil organic matter doesn't feed plants. It doesn't directly, of course. Plants get their carbon from the atmosphere via photosynthesis. But organic matter indirectly feeds the soil life that we now know plays an essential role in plant nutrition and plant health. Oddly enough, the potential to restore life to soil comes down to how we view dead stuff and invisible things—organic matter and microbes.

3

ROOTS OF THE UNDERGROUND ECONOMY

To be a successful farmer one must first know the nature of the soil.

—Xenophon

For much of human history, the role of organic matter in soil fertility was no secret. Farmers and philosophers alike believed that humus—soil organic matter—nourished plants. Until, that is, two key discoveries undercut this long-held belief. The first was the discovery of photosynthesis—that plants obtained their carbon, and thus most of their mass, from the air and not the soil. The second was the observation that most humus was insoluble and could not be sucked up by plant roots. So soil organic matter didn't feed plants.

What replaced the humus theory was the idea of the soil as a chemical reservoir from which plants drew sustenance. In the first half of the nineteenth century, German chemist Justus von Liebig demonstrated that a lack of availability of key nutrients could limit plant growth. He also established that adding elements in relatively short supply dramatically boosted plant growth. Farmers working degraded fields found that adding calcium, phosphorus, or potassium could bring crop yields back up to levels not seen since their grandfather's day.

So did the addition of nitrate- and phosphate-rich guano, bird droppings that were aggressively mined from South Pacific islands. As the supply began to run low in the late nineteenth century, European and North American crop yields were threatened due to centuries of soil loss and degradation. Getting enough nitrogen to crops became a top priority.

Nitrogen gas (N_2) bathes our world—it makes up almost 80 percent of Earth's atmosphere. So you might think that plants could grab all the nitrogen they need from the air. That's what they do for carbon through photosynthesis. But the triple bond between the two atoms in a molecule of nitrogen gas is incredibly stable. So while nitrogen is essential for making amino acids, proteins, and DNA, not much of it is biologically available. This means that nitrogen is often the limiting element for plant growth—especially in soils with little organic matter.

But there was another strategic motivation for securing a steady supply of nitrogen: it was essential for making high explosives. In 1909, a pair of German chemists—Fritz Haber and Carl Bosch—figured out how to synthesize ammonia (NH_3). Using hydrogen gas as a feedstock, they developed a high-pressure, catalyst-based process that worked at high temperatures. The ability to manufacture nitrogen both prolonged the nightmare of the First World War and produced the miracle of cheap fertilizers that could boost crop yields on degraded land, of which there was no shortage.

After the war the Allies demanded the secret to the Haber-Bosch process so they could modernize their own munitions factories. Decades later, after the Second World War, the same Allied countries converted idled munitions factories to fertilizer production, a change that could have been quickly reversed had the Cold War heated up. The widespread availability of cheap fertilizers, coupled with the new fertilizer-loving wheat and rice varieties of the Green Revolution, doubled global crop production.

Although fertilizers can quickly boost crop yields on degraded land, the increase in returns on rich, fertile soil is marginal at best. And as it is, only about half the nitrogen applied as fertilizer gets taken up by crops. The part that isn't doesn't stick around, causing problems off the farm. Most chemical fertilizers readily leach into groundwater because they are soluble by design. So add a lot in the fall and by spring much of it can end up in a river, reservoir, or water well.

THE LAW OF RETURN

Soon after Haber and Bosch uncorked the nitrogen genie, an observant English agronomist began to question the new agricultural gospel. Based on years of work developing large-scale composting methods for commercial plantations in India, Sir Albert Howard proposed his Law of Return in the 1930s to explain why returning organic matter to the fields was essential to soil health, healthy crops, and bountiful harvests. At a time when there was little knowledge of how nutrients reached plants, Howard thought that mycorrhizal fungi played a big role.

In his experience, well-made compost boosted growth of mycorrhizal fungi. And fields with abundant mycorrhizae consistently produced abundant healthy crops. This led Howard to see fungi as nature's recyclers. He suspected that mycorrhizal fungi fed on decaying organic matter and served as root extensions that provided essential nutrients to plants. In Howard's view, chemical fertilizers could not replace soil organic matter, because adding a few elements could never provide all the mineral nutrients and substances in soil that fungi rounded up and delivered to plants.

While Howard grasped the general pattern, he could not really explain why fungi helped nourish plants. To agronomists, Howard's talk of altruistic fungal magic seemed just that. Still, he was sure that the agrochemical bandwagon was speeding down a deadend street.

There is a growing conviction that the increase in plant and animal diseases is somehow connected with the use of artificials. In the old days of mixed farming, the spraying machine was unknown, the toll taken by troubles like foot-and-mouth disease was insignificant compared with what it is now. The clue to all these differences—the mycorrihizal association—has been there all the time. It was not realized because the experiment stations have . . . [thought] only of soil nutrients and have forgotten to look at the way the plant and the soil come into gear.[6]

Howard's lack of an explanation for exactly *how* fungi and other microbes helped plants undermined the scientific community's interest in his challenge to conventional wisdom. Besides, the clear evidence of the near-miraculous effects of fertilizers in reviving flagging crop yields on degraded fields spoke for itself. In short order, Howard's ideas were eclipsed by the Green Revolution's fertilizer-intensive approach to boosting crop yields.

LIFE OF THE SOIL

Another influential agricultural myth is an innocent half-truth I learned in college—that chemistry and physics govern soil fertility. In particular, I was taught that a soil's fertility lay in its cation exchange capacity—its capacity to hold positively charged ions, essential nutrients like potassium (K^+) and calcium (Ca^{2+}), loosely enough for soil water to take them up. This isn't wrong, there's just more to the story.

When farmers send samples off to a commercial lab to find out what's in their soil, it's with an eye toward what they need to add to boost plant growth. But the standard soil chemistry tests only measure the soluble fraction of what is in the soil, the stuff that water percolating through the soil can readily pick up and hand off to plants.

Nutrients tightly held in soil organic matter don't show up in conventional soil tests. Neither do all the nutrients locked up in slow-to-dissolve minerals. At any one time just a fraction of the elements in a soil is in an exchangeable, soluble form plants can take up. So standard soil chemistry lab reports are missing something big: the potential for soil life to convert nutrients from mineral soil and organic matter into forms plants can use. Since the 1980s, advances in soil ecology and microbiology have radically changed our understanding of how microbial life and organic matter interact to govern nutrient cycling and influence soil fertility.

This wouldn't have surprised Sir Albert Howard, or the philosopher-farmers who founded this country. And it shouldn't surprise farmers today. Good farmers may not know all the details behind what makes for fertile soil, but they know it when they touch and see it. I've seen how they pick up soil and rub it between their fingers, asking them-selves: Is it crumbly or dusty, slick or firm? Does it aggregate and hold together, or disintegrate into dust at the touch? And above all, how much organic matter does it hold?

In a way, it's easy to see whether soil is healthy or degraded. The darker the soil the more organic matter—and carbon—it contains. Several generations ago, the amount of organic matter in the soil set the price of agricultural land. Every farmer knew that soil rich in organic matter was more fertile. So did their bankers.

You might think of healthy soil as a particular mix of soil organisms, organic matter, and minerals that forms a thin skin on our planet, like a grand version of lichen coating an alpine boulder. Part alive and part dead, the average thickness of topsoil ranges from about one to three feet. Soil accounts for a thin sliver of Earth's 4,000-mile radius, but its proportions belie its importance. This delicate blanket of rotten rock is what makes our terrestrial world habitable. As the dynamic frontier between the living world of biology and Earth's rocky bones, soil is the realm in which microbial life recycles the remains of higher life into the raw materials for new life.

The history of life on land is a collaborative tale of plants harvesting solar energy, and microbial life mining and recycling nutrients. The first land plants evolved some 450 million years ago. They had partners right from the start—mycorrhizal fungi that hooked up with their roots.

Like today's plants, the earliest ones periodically shed dead roots and leaves, and eventually died. All that organic matter became food for soil organisms that then mined more nutrients from the mineral soil and recycled the dead stuff back into nutrients for the plants to consume. More plants led to more organic matter, which led to richer, more fertile soil. Soon, and for ages ever since, vegetation covered all but the rockiest, driest, or ice-covered landscapes.

Why was this partnership crucial? Consider where plants get their elemental building blocks. They use solar energy to combine carbon dioxide from air with hydrogen from water to make carbohydrates (sugars). Plants also get their nitrogen either indirectly from the air, with microbial assistance from nitrogen-fixing bacteria living in specialized root nodules, or from nitrates they absorb through their roots. Other elements plants need to make their bodies come from rocks and decaying organic matter. Mycorrhizal fungi and soil-dwelling microbes extract mineral nutrients from soil particles and rock fragments and help break organic matter back down to soluble nutrients that plants can suck up through their roots.

Yet roots are not simply straws. They are two-way streets through which carefully negotiated and orchestrated exchanges occur. Plants release into the soil a variety of carbon-rich molecules they make, and which can account for more than a third of their photosynthetic output. For the most part, these exudates consist of proteins and carbohydrates (sugars) that provide an attractive food source for soil microbes. In this manner, plant roots feed the fungi and bacteria that pull nutrients from the soil—from the crystalline structure of rock fragments and organic matter.

When enough microorganisms are present, root exudates don't last long. Microbes chow down on and assimilate most within hours, modifying and re-releasing them in other forms. In addition, with the help of soil-dwelling bacteria certain mycorrihizal fungi use their thread-thin, root-like hyphae to seek out and scavenge particular biologically valuable elements, like phosphorus, from rocks or decaying organic matter. Then they trade the scavenged elements, now in plant-available form, for root exudates. This sets up an exchange through which both sides benefit from the commerce of the original underground economy.

Likewise, the dead cells that roots slough off last for only a few days before microbes consume and reprocess them. The resulting microbial metabolites include plant-growth-promoting hormones and compounds that bolster plant health or aid in plant defense. Some form stable carbon-rich deposits that, in turn, help structure communities of beneficial bacteria in the rhizosphere, a biologically rich zone around plant roots.

Curiously, rhizosphere-dwelling bacteria are more effective at promoting plant growth once a critical microbial density is reached, triggering a process known as quorum sensing. When enough individual bacteria of the right kind are present, they coordinate the release of compounds that aid in promoting plant growth. But, if the population of soil microbes drops too low, they turn off the tap. In other words, they only work if there are enough of them to make a difference to the plant, which in turn produces a healthy exudate return for the microbes. So by pushing enough exudates out into the soil, the plant can cultivate microbial populations that produce compounds useful to the plant. The complexity and adaptation belowground mirrors that aboveground, as plants recruit and feed specialized communities of bacteria and fungi in relationships every bit as specialized as those between flowers and pollinators.

So where do you think you'd find the most bacteria in soils? Where the food is, of course—around plant roots. And where are the most bacteria-eating protists and nematodes? Also around the roots, where the bacteria are. This is another link in the soil food chain—after saprophytic fungi and bacteria consume organic matter, they become enriched with nutrients. Predatory arthropods, nematodes, and protozoa feast on them, then release those nutrients back into the soil in plant-available forms. Because the excrement of these microscopic predators is rich in nitrogen, phosphorus, and micronutrients, it makes excellent micromanure.

In these ways, soil life makes soil fertile. Major elements like calcium, magnesium, phosphorus, potassium, sodium, and sulfur that plants need to make their bodies, and we need to make ours, ultimately come from rocks via the soil. So do essential trace elements like copper, iodine, manganese, molybdenum, and zinc. At every step along the way microbes are intimately involved in making most mineral-derived elements available to plants. And the more of these microbes that are on the job, the more nutrients that are available to plants.

Most (though not all) soils have enough of most elements to grow healthy plants—as long as those elements are unbound from mineral grains and organic matter and in forms that plants can take up. This is the microbes' job. Microbes facilitate getting a suite of essential micronutrients into plants—things like copper and zinc that we don't tend to think of as nutrients but that healthy plants and people alike need in small quantities. Soil-dwelling microbes work like little chemists to convert nutrients to plant-available forms. But in a soil sparsely populated with life, crucial nutrients remain parked outside of a plant's root zone, like goods on a ship stranded at sea far from port.

A BIOLOGICAL BAZAAR

From bacteria to beetles, soil life forms an underground community that breaks down organic matter, yielding organic by-products and metabolites rich in nitrogen and mineral-derived elements. Soil life also influences the ability of plants to defend themselves—when insects or herbivores graze on foliage, some plants exude compounds that rhizosphere-dwelling microbes metabolize. Plants then use the microbial metabolites to drive away the herbivores. In other words, plants outsource the production of pest repellent to microbes that get paid with root exudates. And when the rhizosphere is full of beneficial microbes, pests and pathogens have a harder time finding a seat at the crowded table.

The slow pace of rock weathering and limited availability of biologically critical elements on Earth's surface means that recycling these elements is essential to growing and sustaining abundant life. Over geologic time, microbially mediated processes refined and built up the stock of ingredients circulating through terrestrial ecosystems. Soil life not only turns the wheel of life, it procures and stores nutrients essential for new life and keeps them from leaching out of the soil.

Heavy fertilizer applications can alter soil microbial communities, make the soil acidic, and harm beneficial microbial life. Although crops have access to nitrogen through the fertilizer, elements that microbes previously converted into usable forms may remain inaccessible when the right soil organisms aren't around to do their job. And when plants get all the macronutrients they need for free from fertilizers, they shut off their expensive exudate faucet, denying the microbes that are left in the rhizosphere a much-needed food source. This turns crops into botanical couch potatoes and helps make degraded farmland dependent on nitrogen fertilizers. It also means that while plants may get certain major elements they need to grow, they lose out on the microbial allies that help procure the mineral micronutrients they need to be healthy and mount a robust defense against pests and pathogens.

More than half a century after Sir Albert Howard first proposed his Law of Return, we finally understand how it works. There is a biological basis for the central role soil organic matter plays in growing healthy crops and sustaining bountiful harvests. Fertility isn't only about chemistry and physics. Soil ecology and nutrient cycling driven by microbial life also matter. So even when standard soil chemistry tests say you need to add fertilizers, the right soil life—if present and abundant—may be able to supply what plants need. Growing evidence shows that synthetic fertilizers work like agricultural steroids, propping up short-term crop yields at the expense of long-term fertility and soil health. Consider fertilizers and agrochemicals as like antibiotics—a godsend if you really need them, but foolish to rely on for regular use. And this, of course, is exactly what we've been doing for decades.

In hindsight, we know that our dependence on the plow and fertilizers to pump up crop yields depleted soil organic matter and disrupted the beneficial fungi that extract crucial micronutrients from rocks and deliver them to crops. When we take out mycorrhizal fungi—eliminating or limiting their role in nutrient acquisition—and compromise microbial roles in pest and pathogen control, we have to replace them with fertilizers and pesticides.

But we can reverse this by cultivating beneficial microbial life. And the key to doing this seems to be practices that build soil organic matter—feed them and they will come. Farming practices that maintain high levels of soil organic matter support the diversity of beneficial soil life that in turn supports plant health. Organic-matter rich soils promote beneficial soil nematodes over plant-parasitic nematodes as well as bacterial communities that suppress pathogens. And it is well established that they are more fertile.

Speaking at farming conferences for the past few years, I met farmers discovering how to rebuild fertile soil. They are showing how highly productive agriculture can cultivate soil fertility, using modern technology to update traditional methods and restore productivity to degraded farmland, while sustaining high yields with decreasing energy and input use. Their experiences challenge the wisdom of currently conventional agronomic practices and prove that farming practices that build soil health can reverse trends millennia in the making.

The key to maintaining soil health lies in the world of soil life, in the microbial cycling and recycling of nutrients from mineral and organic matter. Herein lies the good news. For the short lifespan of microbial life means that restoring life and fertility to the soil—and increasing the productivity of marginal farms—is not only possible, but can happen faster than we ever imagined.

4

THE OLDEST PROBLEM

A nation that destroys its soil, destroys itself.
—Franklin D. Roosevelt

Halfway across the field, we stopped and bailed out of the truck. I shielded my shade-adapted Northwest eyes against the August sun and reached down to pick up a handful of soil. Tan grains of feldspar and quartz tumbled through the sieve of my fingers. There was no organic matter, no life—seen or unseen—to bind the soil together. Oddly, this Kansas cornfield reminded me of a California beach I frequented in my high school days, except that cornstalks littered the ground instead of beer bottles.

It hadn't always been like that. When pioneers first arrived in Kansas and busted up the thick prairie sod, they didn't need fertilizers to ensure good harvests. Back then, all one needed to produce a decent crop was a plow, some rain, and a lot of hard work. But the lifeless khaki soil I held in my hands on that bright summer day bore no resemblance to the fertile dark earth that had greeted settlers heading west.

I had come to the heart of the former Dust Bowl to speak at Guy Swanson's workshop on no-till farming methods. Afterward, Joel McClure, a Kansas farmer with a quiet manner and a big mustache, wanted me to see his farm and the "bad soil" he'd inherited and was now trying to fix.

On our drive along Highway 83 toward Oklahoma in Swanson's Impala, we crossed terrain as flat as I'd ever seen. After passing Garden City, where streetlights famously came on at noon during the Dust Bowl's darkest days, we drove over the dry bed of the Arkansas River and sped past a grazing herd of metal cutout buffalo, turned right at Liberal, Kansas, and took a left onto a sandy dirt road. Swanson pointed out a big field that, up until recently, had been plowed annually since the 1930s. Each year, so much sand and silt would blow off the field that the county had to plow the road to keep it passable. Where a century ago a man on horseback could hide in tall grass, persistent prairie winds now piled sand up in six-foot-high drifts along fence lines.

After McClure invited us into his mobile home for lunch, we drove out to see his fields—and his soil. This anemic beach-sand soil with less than half a percent of organic matter was truly dirt compared to the rich black soil that originally lay beneath the buffalo-grazed prairie. A century and a half of farming, and the soil was shot. And yet, his corn crop was beautiful. A heavy diet of fertilizers kept his lifeless soil productive. As I rubbed some between my thumb and fingers, I could see a bleak future in which the end of cheap oil would mean the end of modern harvests dependent upon industrially produced fertilizers. This field laid bare the challenge of maintaining food production on degraded land and perfectly illustrated an underappreciated problem sweeping America's agricultural regions.

BIG MONEY, BAD SOIL

A clean-shaven guy in his early sixties, Guy Swanson is a real Midwest-style storyteller, a people person—the kind of guy who would call a geologist out of the blue with an invitation to speak to a group of Kansas farmers. His passion for no-till farming goes way back. It runs in his family.

Swanson grew up in rolling hill country near the town of Pullman, in eastern Washington. He says his grandfather helped introduce no-till wheat farming to the Pacific Northwest when he began using horse-drawn seed drills in the 1920s to plant a quick crop of no-till winter wheat after a late September harvest of dry peas. The dry pea stalks were removed to a stationary thresher to keep from plugging up the seed drill, leaving the soil naked with no harvest residue. Those peas helped feed the army during the Second World War. But his dad saw their land go downhill over his lifetime. As a boy, he didn't notice anything wrong; at forty he realized that the soil on the hilltops near his farm was no longer rich black topsoil but diminished red subsoil. And at sixty, he talked nostalgically about what his neighbors could've grown if their fields still had topsoil. Swanson hoped that a new twist to his grandfather's no-till farming methods could build the soil back and turn red hilltops black again.

At his workshop, Swanson didn't hide the fact that he was no friend of the fertilizer industry. They probably don't like him much either. His business is to make sure they sell less fertilizer.

He began his talk by stating, in no uncertain terms, that he believes the companies that control the fertilizer industry are functional monopolies. In the United States, two companies hold all the cards: Koch Fertilizer dominates production of nitrogen fertilizers, and the Mosaic Company (a spinoff of Cargill) is the largest producer of phosphate and potash fertilizers. He said these companies can pretty much charge whatever they want since most farmers now depend on their products. And it's the farmers and the government who pay for the subsidized crop insurance that ensures that the fertilizer companies get paid whatever the harvest.

Heads nodded around the room when Swanson likened fertilizer salesmen to drug dealers. I hadn't expected this particular analogy, given all the gray hair in the room—only a few of the all-male audience looked like they could be under forty. But old or young, they had experienced the cycle of addiction for themselves: the fertilizer industry's business model depends on a lack of nutrients in the soil. Nature developed a different solution to plant nutrition, one that has worked for time immemorial. A large and steadily available nutrient reservoir is ideal to reliably nourish plants. This is exactly what a soil rich with decaying organic matter and nutrient prospecting and recycling microbes provides.

These blue-jeaned and baseball-capped farmers were all stuck in a bind, caught between middlemen with little leeway in who they sell to in the commodity market and little choice in who to buy fertilizer from. Trapped in a system they don't control and within which they have no individual leverage, the prospect of paying less for fertilizer was what had brought them to Guy's workshop. They were looking for a better margin—to cut their input costs and help their bottom line.

So when Guy explained that with no-till his equipment can slash fertilizer and diesel use in half, it got everybody's attention, including mine. These are two of the big-ticket costs in modern farming. Not only are they expensive, but crops take up and use less than half the nitrogen and phosphorus farmers apply. Much of the rest ends up in places we don't want, like water wells, streams, Lake Erie, or the Gulf of Mexico. This also means that crops next year can't use what this year's crop didn't. So farmers have to keep buying fertilizer year after year.

To illustrate the benefits of precision fertilization, Guy ran through the numbers. On an irrigated no-till farm with a corn-corn-wheat-corn-sunflower rotation, his system reduced a $200-per-acre conventional fertilizer bill to less than $100 an acre. And it saved $25 to $60 an acre on dryland fields planted half in wheat and half in corn. If applied to a 10,000-acre farm, this adds up to an annual net savings of more than a quarter-million dollars a year.

If that wasn't convincing enough, Guy showed the results of a field-scale experiment on an irrigated farm by the Spokane River in eastern Washington. A huge circular field was divided in half, with each side receiving different amounts and styles of fertilizer application.

One half received the agronomist-recommended 350 pounds per acre of conventionally applied nitrogen fertilizer. The other got just 120 pounds per acre of nitrogen, applied using Guy's precision application equipment. Photographs of both sides of the field showed indistinguishable swaths of mature crops. Guy's system had reduced nitrogen use, and the fertilizer bill, by almost two-thirds—with no loss in yield.

Precision fertilization systems, like Guy's, place a little fertilizer right when and where it's needed most. For his system to work, it has to be used with no-till planting equipment, which, as a bonus, requires about half as much diesel as conventional plowing. By opening a narrow slot in the soil, dropping in seeds, and covering them with earth, no-till farming leaves crop residue at the ground surface, where soil microbes can gradually break it down and convert the nutrients into soluble forms that crops can take up and use. Planting and fertilizing in a single pass or two, farmers save a good deal of time and money.

Guy's simple recipe makes sense: don't till and use less fertilizer by applying only as much as needed, precisely where needed. The combined approach of no-till methods and precision nutrient management helps keep the topsoil and the costly nitrogen and phosphate that farmers do apply where they want them—in the fields. It also helps keep up to four rain-fed or irrigated inches of moisture in the soil, which will typically translate into 28 more bushels of winter wheat an acre.

BELOW THE SURFACE

You can't tell anything is amiss in the fields of western Kansas from an airplane window. You have to look belowground for evidence of lost topsoil and organic matter. And according to Guy, virgin prairie soils were not acidic, but they are now. A radical transition that didn't attract much attention, at least until recently.

In native prairie, most of the biomass lay belowground. The buffalo grazed on vegetation aboveground, but never in one spot for too long.

The great herds would mow the grass down, taking nutrients in and leaving manure behind, fertilizing as they went and returning only after the grass grew back. And grow back it did, with a vengeance, both above and below. The secret to such resilience lay in the subterranean storehouses of carbon stockpiled in plant roots, soil organic matter, and soil life. The thick meters-long roots of prairie grasses didn't just hold the soil in place; they provided a reservoir of energy. The manure and trampled organic matter left in the buffalo's wake recharged the subterranean batteries that powered regrowth.

Fast-forward to the time of the settlers, who tried their hand at taming the land and killed off the buffalo. Plowing up the plains exposed the soil to erosion by wind and rain and accelerated the decay of organic matter. The combination of losing the buffalo and the grass they ate drained the landscape's rechargable batteries.

The green-and-brown agricultural quilt you see out of an airplane window when flying over the American Midwest gives no hint that the land between Denver, Dallas, and Detroit once hosted North American versions of the Amazon and Serengeti. There's not much left of that original landscape in the region's wall-to-wall patchwork of farms broken only by the natural arteries of rivers, and the economic lifelines of highways stringing towns together like beads on an interstate necklace.

Yet there is a telling geometry to the modern landscape. West of the 100th meridian, which runs down the middle of the Dakotas, water scarcity limits crop growth and sweeping arcs of irrigation inscribe great circles on the land. To the east, a rectangular grid memorializes Thomas Jefferson's public land surveys with individual patches of color blocking out square-mile sections, and the 160-acre homesteads of the later Homestead Act of 1862. This patchwork pattern not only testifies to generations of hard work, it structures the foundation of American agricultural wealth.

Hidden from view, however, is the dirty secret of what's been lost since Jefferson's time. One way to understand this is to visit pioneer cemeteries between Minnesota, Ohio, and Missouri. Some of these unplowed islands stand a few feet closer to the heavens than the surrounding fields. But it's not that they have risen. The surrounding land has fallen. Half the native topsoil of Iowa has already started the long trip downriver toward New Orleans and the mouth of the Mississippi. Perennially tilled land lost ground as generations of farmers mined a living from the soil. And though soil loss may be hard to notice year to year, it can really add up over a few lifetimes.

Now the original blanket of famously fertile, rich black earth described in the journals of early pioneers can only be found in history books and scraps of native prairie. Driving through Indiana several summers after my trip through Kansas, I noticed hilltop exposures of tan subsoil rising above black valley bottoms. The pattern told the story: all the topsoil had been plowed off the high points, and piled up in the low points.

Big changes have occurred belowground too. America's soils have already lost about half their organic matter—half the carbon they originally held. Since colonial times, the average amount of soil carbon held in North American agricultural soils dropped from around 6 percent to below 3 percent. The loss of more than half of our soil organic matter happened too slowly for most to notice, but its impact is evident in how reliant on a few major fertilizers farmers have become over the past century.

This happened through the confluence of two things: mechanized plowing and the widespread adoption of synthetic fertilizers. When soil is upended and meets the air, the organic matter in it decays faster, releasing carbon dioxide (CO_2). By 1980, roughly a third of the carbon humanity had already added to the atmosphere since the Industrial Revolution came from plowing up the world's soils, primarily in the Great Plains, Eastern Europe, and China. Overapplication and overreliance on nitrogen fertilizers accelerated the loss of soil organic matter.

Why does this historic chain of events matter today? Soil organic matter feeds the microbial life that helps make and keep soil fertile. In losing half our soil organic matter from agricultural lands, both globally and across the Midwest, we've lost half the fuel for the microbial fires that forge fertile soil. For decades, a ready supply of cheap fertilizers maintained crop yields and masked the effects of tillage-based monocultures on the soil itself. But now that we face a new era, where energy-intensive fertilizers are neither economically or environmentally cheap, humanity can use all the native fertility we can muster.

REPEATING OURSELVES

Globally, soil loss from plowed fields averages a bit more than a millimeter a year. This sounds like a pretty slow pace, until you consider that the average pace of soil production is around a hundred times slower. The net loss pencils out to an inch of fertile soil every few decades. At this pace, it would take just a few centuries to lose all the topsoil from most hillsides.

It may help put this in perspective to think of soil fertility and soil organic matter as natural capital, of fertile soil as nature's bank account. And just as surely as someone who continues to spend more than they make will go broke, a society that runs down its stock of soil or soil organic matter is draining its agricultural bank account and undermining future prosperity.

The idea that soil loss and degradation can influence human societies is not new. It goes back at least as far as the Classical Greek philosopher Plato. In one of his dialogues, he blamed erosion of upland soils for the large deposits of silt that were building deltas out into the sea at the mouths of rivers. Water that used to sink into the ground was running off bare plowed fields and carrying away the soil.

The rich, soft soil has all run away leaving the land nothing but skin and bone. But in those days the damage had not taken place, the hills had high crests, the rocky plain of Phelleus was covered with rich soil, . . . of which there are some traces today.[7]

Plato thought this mattered because it affected the ability of the land to support a large army exempt from agricultural labor. In this sense, he may have been the first to connect how people treat the land to how the land, in turn, will treat their descendants.

Two thousand years later, geologists and archaeologists working together confirmed his story, identifying Bronze Age erosion that stripped soils from Greek hillsides long before Plato's day. While writing *Dirt*, I found an intriguing graph in a book coauthored by a geologist and an archaeologist that charted the population density of Greece's southern Argolid Peninsula from before 5000 B.C. to modern times. It showed rapid rises in population in the Bronze Age, Classical Age, and modern times, separated by two dark ages during which the population crashed. At the peak of each cycle, the population rose to higher levels, something easily explained by advances in agricultural technology. But what explained the roughly 2,000-year periodicity? Could the time it took to erode and then recover the soil account for this boom-and-bust cycle of successive civilizations occupying the same piece of ground?

Intrigued, I started looking at the archaeological record for other societies—and read the four surviving Roman books on agriculture.[8] Spanning a period from the second century B.C. to the first century A.D., they portray how Roman farming changed from diverse poly-cultures on small family farms to commodity-oriented monocultures on large plantations. This sounded all-too familiar, just like what had happened in twentieth-century America.

The big change occurred after the Romans sacked Carthage. They enslaved the survivors and brought them back to central Italy to work in the fields. This influx of slave labor changed the focus of Roman agriculture to large plantations growing single crops. Coupled with the Roman love of the plow, this massive shift in farming practices triggered serious soil erosion across central Italy, as recorded in lake sediments and on the exposed foundations of ancient buildings constructed at known dates. In his epic poem the *Georgics*, the Roman poet Virgil attributed the destruction caused by floods to upland plowing and recommended plowing along contours rather than straight up and down slopes. Although lowland sedimentation piled up to create the famously pestilent Pontine Marshes, enough sediment still made it to the coast to build the shoreline out and strand Rome's ancient port of Ostia several miles inland.

By the peak of the empire, North African farms were essential for feeding Rome. It became a capital crime to delay the grain ships traveling from afar to sustain the heart of the eternal empire. While many of us learned in school that the Romans salted the earth around Carthage after they destroyed their rival city, few know that Roman farmers were back hundreds of year later, forced to work that same ground after they'd inadvertently wrecked the soils of their own homeland. Today, the North African and Syrian hills that once exported grain to feed Rome stand as bare rocky slopes, the local populace saddled with degraded land.

Centuries later, soil erosion began to plague colonial North America, shaping U.S. history in ways I was never taught in school. Tobacco was the crop that got the upstart British colonies onto solid economic footing—it could survive the long trip back to London and arrive in marketable condition. Although quite profitable, the bare-ground style of colonial tobacco cultivation was highly erosive and degraded the soil quickly. A farmer could expect just several years of profitable harvests from a freshly cleared piece of land before needing to move on to fresh soil. Here was soil-driven periodicity on a smaller scale and within a shorter time frame.

In those days, however, land was expensive and labor was cheap in Europe, while the reverse was true in North America. This drove European interest in soil husbandry. But life in the colonies was different, especially in the South. The combination of the erosive effects of tobacco cultivation and the low cost of land fostered the growth of large landholdings—plantations. And to make plantation agriculture profitable required slavery. Soil erosion and land abuse flourished alongside human bondage.

Still, the state of the land concerned some plantation owners. Among them, George Washington and Thomas Jefferson, who, though they may not have agreed on much politically, both viewed the impact of erosive colonial farming practices with alarm. Each tried to improve his land. Jefferson obsessively researched European farming practices and collected seeds and plants throughout his life. On his plantation at Monticello he alternated cover crops with his main crops, planted turnips, vetch, and buckwheat as green dressings, and applied manure to sweeten the soil.

Washington blamed the widespread practice of continually growing nothing but tobacco for wearing out colonial fields. He too experimented with crop rotations and was known for filling in gullies on his land. In a 1796 letter to Alexander Hamilton, Washington worried about what continuing destructive farming practices would do to an agricultural nation.

A few years more of increased sterility will drive the Inhabitants of the Atlantic States westward for support; whereas if they were taught how to improve the old, instead of going in pursuit of new and productive soils, they would make these acres which now scarcely yield them any thing, turn out beneficial to themselves.[9]

Just as Washington feared, degradation of the soils along the eastern seaboard began to push American farmers west in search of fresh land. But Washington didn't realize how this would help tear the country apart half a century later. As an exodus of farmers fled west, the issue of slave labor became a make-or-break economic issue for plantation owners decamping to plow up the virgin soils of new states—and for those who stayed behind.

In the decades before the Civil War, selling slaves to those heading west was big business for the original states of the Southeast. Data from the census of 1860 show that slaves accounted for almost half of southern wealth, including the value of land. Abolition of slavery would devastate the southern economy. The ensuing economic tension helped split the country, with political repercussions that still frame politics today.

How serious was colonial soil erosion in the Southeast? A few years ago colleagues at the University of Vermont measured the long-term background rates of erosion from sampling river sands from ten watersheds stretching from Virginia to Alabama. The concentration of beryllium-10 (^{10}Be), an isotope produced when cosmic rays smash into quartz grains, can be read like a clock to estimate how long it has been since the mineral grain lay shielded at depth below ground. They found that for thousands of years before European colonization, erosion averaged around a millimeter a century. This is a hundred times slower than the estimated average millimeter-a-year pace of colonial upland erosion. An estimated six inches to a foot of topsoil has eroded off the farms across the region during and since colonial times.

I recently saw this landscape shorn of topsoil with my own eyes when I had the opportunity to visit a modern tobacco plantation and several farms in North Carolina. These long-farmed upland soils are really the original subsoil now exposed at the surface. The soil in the tobacco field was crusty khaki sand. When I dug a hole beneath a tall forest right next to the tobacco field, I found the missing topsoil. The contrast was striking. The dark brown soil full of organic matter held together in clumps by mycorrizal fungi was completely different than the neighboring long-farmed soil.

If American agriculture could erode off virtually all of the topsoil across a broad region in just a couple centuries, imagine what the Romans and Greeks did with their far longer run at it. Here we have strong evidence that societies can erode themselves out of prosperity, which of course begs the question of whether we're playing out an old story that we know has a bad ending.

Of course, I'm not the only person since Washington and Jefferson to notice such things. A number of writers have commented on the role of soil erosion in impoverishing whole regions. After all, no one would argue that modern Iraq looks much like the Garden of Eden.

When I dug into the geoarchaeology literature, I found that playing out beneath the demands of rising populations, the vicissitudes of climate, and the ravages of war, erosive farming practices undermined agricultural civilizations time and again throughout history. In many ways, soil degradation set the long-wavelength pattern of history, as wars, natural disasters, and climate shifts pulled the trigger on environmental guns loaded by soil loss and degradation. A common storyline ran through the rise and fall of many civilizations: farming started on flat, well-watered floodplains before growing populations spread their fields onto upland slopes with a thin blanket of soil above bedrock. After centuries of erosion stripped soil from the land, it became harder to sustain large populations, rendering societies vulnerable to disruption.

Since the dawn of agriculture, the cascading effects of soil degradation impoverished the descendants of once prosperous civilizations. Put simply, the slow pace at which nature makes fertile soil ensured devastating consequences for societies that failed to protect their land. For once soil eroded away, nature was in no hurry to put it back. From ancient Greece to modern Haiti, civilizations crumbled beneath agricultural practices that degraded soil fertility and eroded fertile topsoil. Versions of this story played out—to one degree or another—in the Middle East, Classical Greece, Rome, Easter Island, Central America, and early China.

Even the obvious exceptions to those long-lived civilizations located along major rivers help make the general case. The floodplains of the Nile, Indus, Brahmaputra, Tigris, Euphrates, and major rivers of China were all fertilized by frequent deposition of fresh silt eroded off distant uplands. In other words, soil loss in Sudan and Ethiopia subsidized the long-running fertility of Egypt, just as the Himalaya supported India and Tibet helped feed China.

For millennia, we have made a living mining soil. It's understandable that most people take soil for granted—it doesn't blossom before your eyes like a plant or a child. In reality, it forms so slowly that a single storm can strip a century of soil formation off a bare, freshly plowed field. Soil problems play out over such long time periods that their effects manifest as slow-motion disasters. But looking back through history, it's clear that societies that abused their soil through erosive farming practices shortchanged their descendants. And today the twin problems of soil erosion and degradation of soil fertility once again threaten civilization's agricultural foundation.

Driving back from Kansas to Denver, Guy Swanson and I passed field after field with no visible trace of organic matter. We talked about how farmers whose great grandparents plowed rich black prairie soil now work light brown earth, the color starkly revealing the toll of a century of modern farming. The message is all too clear if you know how to read it.

As we sped along U.S. 287 near Eads, Colorado, the wind came up from the south, and dirt clouds rose skyward off freshly plowed fields. It was only 4 P.M., but the semi trucks passing by had turned on their lights. Dust blew up over the road, out of driveways and off fields, and I rolled my window up as a brown curtain billowed toward us across the plains. Ground-level haze collected into wispy fingers pointing skyward, like despairing ghosts of the Dust Bowl.

But no dirt was blowing up from the grassland along the road. Nor from the occasional crop-stubble-covered field we passed. It was clear that while the wind could entrain the soil if given half a chance, plant cover could lock it down. This is no secret. As we approached the Denver airport, a band of no-till farms were doing just that. Tillage was banned around the airport in order to cut down on dust storms.

As he pointed this out, Guy noted that government programs that subsidize tillage and crop insurance inadvertently promote soil loss, eroding our economic foundation and putting nitrates into water supplies. The modern no-till farmers of the former Dust Bowl region I talked to saw how "power farming" ruined their soil—taking the organic matter down to less than a percent or two, turning rich prairie soils into the functional equivalent of beach sand, and leaving farmers dependent on fertilized fertility.

The similarities between this modern story of degraded soil in

Kansas and those of ancient societies set me off on a search for viable solutions to the problem of soil degradation. What could be done to avoid the fate of civilizations that neglected to safeguard the fertility of the land that fed them? In pursuit of this question, I soon confirmed that the first step involved retiring one of our most powerful tools—the plow.

5

DITCHING THE PLOW

*The plowshare may well have destroyed more options for
future generations than the sword.*

—Wes Jackson

A couple of years after visiting Guy Swanson in Kansas, I accepted
an invitation to address the 2014 World Congress on Conservation
Agriculture in Winnipeg, Manitoba. I gave one of my *Dirt* talks and
figured I'd get the usual questions. Instead, I encountered something
unexpected—heated arguments between conventional and organic
farmers about the best practices for restoring soil.

In my presentation, I advocated combining no-till and organic
practices as a recipe for preserving the soil and its fertility over the
long run. I knew from my research and travels around the American
Midwest that no-till methods were needed to reduce erosion, and I
thought that organic methods had the best shot at maintaining fertility in a postoil world. But while fielding questions afterward it became
clear to me that although everybody in the audience endorsed no-till
farming as a way to restore degraded soil, they were bitterly divided
over whether to do it organically or with agrochemicals and genetically modified crops.

Passions flared as conventional farmers squared off against their organic brethren. A charismatic Australian fellow claimed organic no-till didn't work down under, and that he had to use herbicides and fertilizers to grow his crops. Guys from small-town Pennsylvania and Europe rose to swear organic no-till not only worked really well, it could work *anywhere* and prove cost-effective to boot. It was not lost on me that, despite these disagreements, they all agreed on one thing: ditching the plow.

I came away from the conference with a new appreciation for the way the philosophy underpinning conservation agriculture bridges different practices and perspectives. All too often, it seems, conflict gets characterized as between organic Luddites and genetic engineering automatons. While such posturing can rally partisans, it also distracts us from the potential for progress. Such as how the young science of microbial ecology is providing insights into what really matters—soil health.

At lunch the day after my talk, Howard G. Buffett, the farmer-son of financial wizard Warren Buffett, shared his vision of what he calls the Brown Revolution. With grace and humor, he highlighted inspiring stories of farmers restoring fertility and profitability to their farms. He wasn't defending ideological turf; he neither argued for organic practices nor pushed biotechnology. Instead, he argued for adopting practices that built soil fertility and thereby reduced a farmer's need—and expenses—for agrochemicals.

I was intrigued by this perspective. Buffett's Brown Revolution sounded like fertile middle ground between organic and agrochemical farming. But how well did it really work? As I looked into it, I began to realize that the principles of conservation agriculture could dramatically and rapidly reform conventional agriculture. Could putting soil health at the center of every type of farming offer a path to real progress that was both rapid and profitable?

A NEW WAY

Conservation agriculture is a system of farming that rests on three simple principles: (1) minimum disturbance of the soil; (2) growing cover crops and retaining crop residue so that the soil is *always* covered; and (3) use of diverse crop rotations. These principles can be applied anywhere, on organic or conventional farms, with or without genetically modified crops.[10]

Despite the heated debate among proponents who have different ideas about how to implement the three fundamental practices, all sides agree on one thing. We need farming methods that are far less harmful to beneficial soil life. For soil health is built upon the backs of soil organisms, from the common earthworm to specialized bacteria, mycorrhizal fungi, and other microorganisms. At its heart this is what conservation agriculture is all about—practices that promote and protect the tiny multitudes that we now know help crops grow and keep soil fertile. It's worth a few words on how conservation agriculture's three principles accomplish this.

Daring to question the plow may seem heretical, but using one guarantees major-league soil disturbance. That's why going no-till is at the heart of conservation agriculture. It leaves nonharvestable plant parts—crop residues—as a soil cover. This means that, after a crop is harvested, plant remains, whether cornstalks or wheat stems, are not removed or burned. They decompose in the fields where they form a carpet of organic matter at the ground surface—mulch. Soil microbial biomass increases rapidly after conversion to no-till. So do soil fauna. Mulched plots have higher bacteria, fungi, earthworm, and nematode populations. Frequent tillage, in contrast, reduces soil microbial biomass and disrupts mycorrhizal hyphae that help deliver phosphorus to plants, among other negative impacts.

Cover crops planted in seasons between commercial crops, and that

are either mowed down or killed before or during subsequent plant-
ings, help suppress perennial weeds and return nutrients to the soil as
they decay. Covering the ground also promotes aboveground biomass
and biodiversity, particularly beneficial insects that help keep pests in
check. Crop rotations also help prevent insect pests and plant pathogens from gaining a toehold. Complex rotations that vary the sequence
of cash crops and cover crops deny pests and pathogens opportunities
to take hold and help break up cycles of insect pests and crop diseases.
This, in turn, helps reduce the need to use conventional pesticides.

The benefits of increased activity and diversity of soil life include
increased water infiltration and soil organic matter, which improve
runoff water quality and soil structure. Soils with high microbial diversity are also a tough place for pathogens to take hold and persist. This
translates into healthier plants. Fewer crop diseases strike, and if they
do, they are not so devastating. Crop rotations also increase microbial diversity, reducing the risk that pests and pathogens will dominate a soil ecosystem. The net effect of adopting all three conservation
agriculture elements is to maintain or increase crop yields (to varying
degrees depending on, among other things, the initial state of the soil)
and decrease fuel, fertilizer, and pesticide use. Conservation agriculture also takes less labor than conventional tillage. This translates into
substantial savings for farmers because they spend less on inputs.

This set of farming principles turns contemporary conventional
farming wisdom on its head. The now-conventional view considers
plowing essential to control weeds, that erosion is an unavoidable
result of rainfall, that cover crops and crop rotations are optional, and
that chemical pest control is a necessity. Under conservation agriculture plowing is avoided and crop residues are left on fields as mulch to
prevent erosion, cover crops and crop rotations are essential rather
than optional, and biological pest control becomes a practical and
effective option.

Conventional tillage does provide weed control, but under conservation agriculture herbicides or cover crops accomplish the same thing

and dramatically reduce soil erosion. Mulch protects the soil surface from raindrop impacts and erosion by overland flow, and no-till farming reduces topsoil erosion as much as fiftyfold. Nitrogen-fixing microbes, legumes, and cover crops that build up soil organic matter can largely, if not completely, replace conventional fertilizers.

The three parts of conservation agriculture work together as a system, but farmers don't always adopt them all. Results are highly variable for partial adoption and a number of review studies report mixed effects on yields and soil organic matter from adoption of no-till. While few of the studies in such reviews addressed outcomes of adopting *all three* system components, the ideas behind this new system have deep roots.

Japanese farmer Masanobu Fukuoka was on to conservation agriculture in the late 1970s. His book, *The One-Straw Revolution* urged the organic agriculture movement to incorporate several ancient practices. He described planting directly into prior crop stubble at harvest time as farming in nature's image. Fukuoka challenged farmers to treat the sequence of annual activities on a farm as an integrated system of applied ecology. His recipe for plentiful harvests with less work was to cooperate with nature and schedule planting and harvesting so that each crop would set the stage for the next. This kept the fields occupied by desired plants, denying weeds the chance to get started in the first place.

Both the emerging conservation agriculture movement and Fukuoka's philosophy embrace a set of principles that, if put into practice, can improve soil quality and boost soil health—and crop yields. When I first learned of them, I was struck by how they lead to thinking about farming differently—as feeding the soil so it will in turn feed us. It boils down to investing in fertility through enlisting trillions of microbial helpers, agriculture's living foundation. Proponents argue that adopting all three principles of conservation agriculture, and adapting them to the circumstances of individual regions and farms, can transform soil-mining farms into soil-building ones. Imagine if the farmers of the Roman Empire or biblical times knew to employ these methods. How might that have changed the course of human history? Perhaps the answer to one of our longest-running problems is really pretty simple after all.

THE ROAD TO THE DUST BOWL

For centuries, the plow has been used to control weeds, prepare the seedbed for planting, and mix fertilizers (chemical or manure) into the soil. Tillage promotes uniform seed germination, and can allow the crop to germinate before weeds and grow without competition. Tillage also exposes organic matter to air, spurring decay that releases nutrients that promote crop growth. These effects benefit the farmer over the short run, but result in a long-term cost to the fertility of the soil, through soil erosion and accelerated degradation of organic matter.

Over time, farmers and the rest of us came to associate tillage with good farming, and to see the plow as a revered agricultural icon. Farmers loved the burst of fertility that came from plowing. But they overlooked or were unaware of the side effects—soil erosion and loss of soil organic matter, soil structure, and soil life—all of which degraded soil fertility over time.

The earliest "farming" by hunter-gatherer societies used no-till-like methods, scattering seeds on the ground or placing them in shallow holes. A farmer could use a forked branch or digging stick to scratch a hole into the fine silt of a floodplain or delta soil, then drop the seeds in. It is no coincidence that these are the environments where early farming—and the earliest civilizations—took root.

The plow evolved from simple tillage tools pulled by animals, used between 5000 and 3000 B.C., to the iron Roman plow, the forerunner of our modern plow. The soil-inverting plow arose about a thousand years later, at the close of the first millennium, and in 1784, Thomas Jefferson designed the moldboard plow that graces the seal of the U.S. Department of Agriculture (USDA). Jefferson's plow moved so readily through the soil that the French Society of Agriculture awarded him their gold medal. Decades later, in the 1830s, a blacksmith by the name of John Deere began marketing a cast-iron plow based on Jefferson's design. Readily pulled by a pair of horses or mules, Deere's plow opened the Midwest to homesteaders, becoming affectionately known as the Prairie Breaker.

Decades later, high prices for grain exports to Europe during the First World War enticed American farmers to plow up everything they could. Mechanized versions of Deere's handiwork opened up the semiarid prairie in the heart of the continent. Then the drought hit. The years of 1934 to 1936 were extremely bad—rainfall was less than half the normal amount.

Though drought may have triggered it, the Dust Bowl was a manmade disaster. For thousands of years, comparable droughts occurred in the Great Plains every few centuries, without resulting in wholesale soil erosion. But when farmers plowed up the plains, the soil lost its living anchor of deep-rooted prairie plants. The land fell apart under high winds when the next major drought hit.

This didn't come as a complete surprise. Throughout the 1920s, soil conservationist Hugh Hammond Bennett labeled soil erosion a "national menace." He preached the gospel of soil husbandry, of caring for the soil as an intergenerational trust. In September 1933, President Roosevelt appointed Bennett director of the newly established Soil Erosion Service. A natural showman with astute political instincts, Bennett didn't waste opportunities to marshal support for soil conservation. And the Dust Bowl gave him plenty to work with.

He'd been in his new position less than a year when a towering three-mile-high cloud of dust blocked out the sun from Texas to the Dakotas to Ohio. Twelve million tons of dirt rained down on Chicago. It was no way to herald spring. A short time later, on May 12, dust descended on the nation's capital. A dark cloud drifted through the Oval Office windows, covering Roosevelt's desk with midwestern dirt—surely capturing his attention.

Tipped off by field agents who would call in to report on a dust storm's progress across the country, Bennett planned strategically. On March 6 and 21 of 1935, great black clouds of dust once again blew through Washington, just in time for congressional hearings on soil conservation legislation. And on April 2, as Bennett testified before the Senate Public Lands Committee, leaden shadows filled the sky and blocked out the daylight, providing Bennett with the perfect dramatic visual to accompany his call to save America's soil.

Several weeks later, President Roosevelt signed the Soil Conservation Act, establishing the Soil Conservation Service.[11] Bennett was appointed chief of the new agency. Two years later, Roosevelt authorized formation of soil conservation districts to serve as the front line in the battle against erosion.

The national disaster of the Dust Bowl also prompted heated debate about plowing. Initially, no-till proponents ran into deeply entrenched skepticism—and public controversy. But eventually their success at reducing both soil erosion and farmers' input costs fueled an ongoing shift to no-till practices, used today on roughly a third of U.S. cropland.

Edward Faulkner was an early leader in the no-till farming movement. His 1943 book *Plowman's Folly* contended that plowing was unnecessary and counterproductive over the long run. He advocated planting directly into the surficial layer of organic matter, where plants naturally germinate when they fall to the ground. After several decades working as a county agricultural agent in Kentucky and Ohio, he concluded that not only was there no good reason to plow, it did more harm than good.

He'd come to this heretical belief much the way I had—through spending time in the backyard. After the crash of 1929, Faulkner tried to grow corn in his brick-like soil. Throughout the Depression, he dug organic matter into the ground, mixing leaves into the uppermost six to eight inches of soil to mimic the effects of plowing. He wasn't impressed with the results. So, in the fall of 1937, he tried a new tactic and simply left the leaves on the ground surface to mulch by default, instead of expending the effort to till.

The next spring, his previously stiff clay soil could be raked like sand, its texture granular instead of brick-like. He grew his best crops yet, only this time without any fertilizer and with hardly any watering. All he did was weed.

Faulkner was excited by these results, but he knew that an anecdotal challenge to conventional wisdom wouldn't convert skeptics. Extension agents, farm papers, and almost every source of farming advice encouraged farmers to plow deep down into the subsoil. He wasn't completely surprised that Soil Conservation Service agents didn't see the relevance of his backyard experiment to full-scale farming.

So Faulkner leased a field to see for himself if planting through a ground cover of chopped-up crop stubble would work on a larger area, the same way it had in his backyard garden. Sure that Faulkner was a fool to forgo plowing, fertilizers, and pesticides, neighboring farmers were shocked when the madman's yields exceeded their own for several years in a row. Faulkner told farmers to stop tilling and start leaving organic matter on the ground. He sparked further controversy when he argued that fertilizers simply decreased "the devastating effects of plowing."[12]

In 1946, several years after Faulkner's book was published, Walter Thomas Jack published a rebuttal of sorts, titled *The Furrow and Us*. In it, Jack argued that farmers around the world found that plowing increased the fertility of their land. Bickering between Faulkner and Jack spilled over into the popular press. It even made it to the pages of *Time* magazine, which characterized their debate as the "hottest farming argument since the tractor first challenged the horse."[13]

Who was right? To some degree, both were. We now know that tillage accelerates decomposition of soil organic matter. What this means, of course, is that plowing gives crops a nutrient boost in the short run. But, if the nutrients aren't replaced, it runs fertility down over the long term. The development of powerful tractors compounded this situation. Mechanization eliminated the need for on-farm pasture and forage, displacing draft animals and reducing the return of their life-giving manure to the land.

Production of direct-seeding machinery in the 1940s allowed planting without tilling, but the lack of herbicides for weed control handed the first round of the debate to Jack and the plow proponents. The development of the herbicide 2,4-D after the Second World War, and paraquat a couple decades later, increased interest in various forms of no-till farming. With chemical herbicides replacing tillage as a primary means of weed control, low-till and no-till methods began to catch on.

No-till farming proved attractive once coupled with easy weed suppression. It cut erosion down enough to keep the soil in place, even under intensive cropping. Retaining more ground cover on fields helped rainwater sink into the ground, which helped crops weather droughts. And no-till reduced diesel bills due to fewer passes of machinery across the fields. Eventually, adopters that left crop stubble on their fields found that outlays for fertilizer dropped too, as soil organic matter gradually rebuilt soil fertility.

RELEARNING OLD LESSONS

I learned that there's more to this story of no-till farming and conservation agriculture when I ran into Rattan Lal again at a conference in Malmö, Sweden, in the spring of 2015. Lal is thin and soft-spoken, with big glasses, a self-effacing manner, and a sparkling laugh. I figured that he probably has more experience than anyone in terms of conservation agriculture in the tropics, so while waiting to meet our hosts in a hotel lobby, I asked him how he came to focus on agriculture.

Lal explained that he grew up on a small farm in Punjab, in what is now the Indian state of Haryana, after his family had fled from Pakistan upon partition in 1947. They'd left behind a 9-acre farm, so when his father received only an acre and a half, he thought the village revenue officer who'd evaluated their land had cheated them. But years later, Lal realized that his family actually got a good deal, exchanging an old farm with a high water table and salinity problems for high-quality irrigated land. By then he knew the quality of the soil was as important as the size of a farm.

A lifelong vegetarian, Lal looked after cattle as a child. His mother passed away when he was two, and he often went to live with his aunt in town. After completing, and excelling in, the village high school in 1959, he gained admission to Punjab Agricultural University, where he landed a scholarship for Pakistan refugees that paid 12 rupees a month (about $1 a week). When he completed college in 1963, knowing how to plow a straight furrow and identify weeds, his farm community was not impressed. "It was a big joke," he said. "This child we sent to college has been learning how to plow!" Yet the experience had lifelong impact, setting Lal on the path to question why farmers plowed in the first place.

Having earned the highest grades in his graduating class, a fertilizer company offered him a 100-rupee scholarship to continue his studies. But his professor told him, "Don't you sign that damn agreement—it will make you work for them for years. I'll send you to Ohio State." Lal was admitted in 1965, just as no-till began attracting renewed academic interest.

He found himself at the school where Faulkner and Jack had argued over plowing twenty years before. Most farmers still habitually relied on the plow for weed control, yet Lal couldn't see why farmers should risk all the erosion tillage can cause when better ways to put seeds into the ground and manage weeds were available.

He completed a soil science Ph.D. in just two and a half years, graduating in September of 1968 at the ripe old age of twenty-four. Taking an appointment as a research fellow in Australia at the University of Sydney, he landed in a different school of thought amid different soils. There he worked on vertisols, soils that expand and contract enough upon wetting and drying to produce big cracks every dry season. Lal wanted to know why these soils produced so much runoff and erosion when plowed, despite the expectation that rain would just sink down into the cracks.

Through a series of clever experiments, he found that when the soil dried out it became impermeable because the heat produced from water condensing within the soil during the process of infiltration caused the soil to break into tiny pieces that then quickly sealed up the ground surface in the next rainfall. This promoted rapid, highly erosive runoff instead of allowing water to sink into the earth where plants could drink it in.

The solution proved remarkably simple: keep the soil covered in mulch so it didn't heat up so much and dry out. Time and again, over the next several decades, Lal's research led him back to the value of protecting the land surface with a cover of living plants and organic matter—nature's recipe for sustained fertility.

The next chapter in Lal's journey started when Dennis Greenland, a professor from the University of Reading, came to visit the University of Sydney and left impressed with Lal's work. After Lal accepted a job at the newly established International Institute of Tropical Agriculture in Ibadan, Nigeria, Greenland was appointed director of research at the institute.

By early 1970, Lal started to investigate solutions for the problems of subsistence farmers. The age-old practice of clearing a patch of forest, farming it for a few years, and then abandoning it back to the forest for decades allowed soil fertility to rebuild during long fallow periods. But population pressure had increased to the point where most of the land was under more or less continuous cultivation. And while the Green Revolution had increased yields on large farms, subsistence farmers lacked the means to buy expensive fertilizers and agrochemicals. They needed something else.

With a visit from the board of trustees looming, Lal helped the staff to hurriedly clear forest and plant their first crops. But the day before the bigwigs arrived, intense rainfall washed out all the freshly cleared plots, leaving horribly gullied fields in its wake. The disaster convinced Lal that Western-style plowing and continuous cultivation was a serious mistake for the fragile soils, intense rains, and resource-poor smallholder farmers of the tropics.

His new boss agreed. The only soil scientist in the Royal Academy at the time, Greenland had seen this before. In the 1950s, the British government hatched a development scheme to address the postwar shortage of cooking oil in Europe. Sure that Western equipment could outcompete and outproduce traditional farming practices, Parliament spent 25 million pounds in Tanganyika (modern Tanzania) to clear 150,000 acres of forest and plant peanuts to make peanut oil. They recruited veterans into a "Groundnut Army," equipped with surplus U.S. Army tractors and odd half-tractor, half-Sherman-tank hybrids. The campaign proved disastrous.

They had little idea how erosive the region's intense rainfall could be on freshly disturbed soils. To clear the whole area in one go, they dragged ships' anchor chains, stretched between tractor-tanks, across the land to clear the vegetation. In short order, intense storms washed away much of the topsoil. In the next hot season, the now-bare fields baked, ruining the peanut harvest. While the project caused a scandal in Parliament, it was not unique. In Nigeria, Ghana, and across the tropics, bulldoze-and-plant schemes ended in similar fashion when predictably savage rainfall ravaged freshly plowed fields.

The failure of Lal's first plots seemed all too familiar. "Rattan," Greenland exclaimed, "it's the Groundnut Affair all over again!" With support from the Rockefeller and Ford foundations, they set up a series of experiments to investigate where well-intentioned Western farming efforts fell apart.

Conventional wisdom at the time held that the degree of the slope was the primary control on erosion. Everyone knew that steeper slopes eroded faster. Lal's data confirmed that slope was indeed the primary driver of erosion *if* the ground was bare, as it was after plowing. But he also found that steep slopes didn't erode if covered with mulch. Lal's experiments showed that the biggest influence on erosion was whether or not you left mulch on the fields after harvest.

At the time, West African governments subsidized plowing schemes. Farmers and agronomists believed that plowing helped water sink into the ground instead of running off over the surface and causing erosion. But Lal found the opposite was true. Less runoff—and therefore less erosion—occurred on plots that were not plowed. Leaving more than 4 tons of plant residue per hectare completely eliminated erosion on all but the steepest slopes. And after an initial year to establish the ground cover, no-till methods proved just as effective as heavy mulch. Where crop residue stayed on the fields earthworms were abundant and the water sank into the ground instead of running off. Plowing wasn't the solution—it was the problem.

As in Australia, another major effect of clearing and plowing was that bare ground heated up when exposed to the sun. In Nigeria, plowed fields had soil temperatures more than 20 degrees higher than adjacent forest soils. But no-till practices kept soil temperatures close to those of forest soil and also retained more water.

Lal was shocked. He knew soil life shut down when soil temperatures rose above about 90°F in plowed fields. And once biologic activity stopped, fertility fell as soil structure declined and erosion ensued. The best thing to do, it seemed, was to keep the ground covered and let the worms and termites do the plowing. But to do that they needed to be fed. And what they ate was organic matter—mulch.

This new view did not find favor among agronomists, however. When apprised of Lal's results, the French chairman of the institute's board said he simply could not believe that Lal would seriously recommend not tilling. What kind of madness was that? The foundation experts and development agency guys—everyone whose opinion mattered—all knew that more plowing was better. So Greenland and Lal were sent off to consult colleagues who could correct their misguided views. But what they found instead convinced them of the validity of their findings all the more.

Particularly informative was a visit to influential scientists who advocated plowing at the end of each rainy season based on experiments comparing tilled and no-till plots. They told Lal that after harvesting, they carefully removed the crop residue from their no-till plots to simulate the grazing practiced by traditional farmers throughout the Sahel. This left no mulch, exposing bare ground. They didn't use herbicides on their no-till plots, so weeds proliferated and crowded out the crops. They then compared the resulting feeble harvests to those from the plowed plots where tillage had suppressed the weeds.

Lal considered this a misleading comparison. Leaving crop residue on the fields was the whole point of not tilling—the reason it worked to stop erosion. And plowing suppressed weeds much like an herbicide did. The agency scientists' rationale for their study design was that African farmers didn't use herbicides and often removed crop residue to use as cooking fuel. But Lal saw that the experiment did not evaluate whether no-till worked if crop residues stayed on the fields—as his own experiments had done. Here was what the agency scientists were missing—their comparison was no comparison at all.

Lal came to the conclusion that resistance to getting no-till adopted in Africa stemmed from the general belief that herbicides were necessary for weed suppression if one didn't plow. And at that time the primary herbicide available was paraquat, which was too expensive for subsistence farmers. That left using cover crops and mulch as the only options to suppress weeds.

Based in part on Lal's experiments, Greenland published a prescient paper in the journal *Science*, titled "Bringing the Green Revolution to the Shifting Cultivator." After Lal mentioned this 1975 article to me, I looked it up and saw that Greenland had spelled out the essential principles of conservation agriculture. He pointed out how the Green Revolution excluded traditional subsistence farmers, who lacked capital and could not afford new equipment, chemicals, and patented seeds. Changes in practices—new ways of thinking—offered subsistence farmers their best shot at continuous cultivation. Farmers needed to eliminate tillage to reduce erosion, and plant a diverse mix of high-yielding crops that included legumes (like cowpeas).

Greenland anticipated criticism of his recommendations, which ran counter to the prevailing view that the path to agricultural development lay through mechanization. To liberate farmers from "the drudgery of the hoe," he recommended the use of herbicides spot-applied by low-cost backpack sprayers. Then he went on to emphasize how proper mulching could be just as effective for weed suppression. Greenland thought that minimizing disturbance of the ground and planting a complex mix of cover crops could not only move subsistence farmers from traditional shifting cultivation to continuous cultivation, but double or triple their yields in the process.

After five years of experiments, Lal and Greenland concluded that what matters most is not what farmers grow but how they grow it. Lal developed guidelines, based on soil moisture and texture, for the most appropriate types of farming for different environments and soils. Eventually, after a decade of work in the tropics, Lal boiled his experience down to two key recommendations: don't clear the forest if you can avoid it, but if you do, make sure you leave the ground covered with vegetation or mulch.

By the time he left Africa in 1987 to take a position back at Ohio State University, Lal had worked on soil problems in fourteen countries on four continents. Although the settings, soils, and climates differed greatly, his experiments all pointed to the value of ground cover and mulch for preventing destructive erosion and for keeping soils fertile. In subsequent publications he emphasized the importance of mulching and cover cropping for sustaining the productivity of tropical soils.

Within two years of his departure from Africa, trees were growing through his experimental plots. The grand experiment was over. He'd figured out something that would work for subsistence farmers. So why were his findings all but ignored?

Funders and aid agencies alike wanted breakthroughs and rapid revolutions, not gradual improvement of the soil. Commercial interests pushed to develop solutions that could be commodified; they wanted agrochemical products, not practices that anyone could adopt for free. No modern, forward-looking foundation or agency wanted to hear about mulching or growing a diversity of crops. Such simple answers did not—and still don't—fit the technophilic narrative of progress.

APPLES TO ORANGES

Though disappointed with the lack of interest in his African experiments, Lal saw his ideas find more fertile ground in South America, where intense tropical rain routinely devastated plowed fields. In 1971, Herbert Bartz began experimenting with reduced tillage on his farm in southern Brazil. It didn't go well at first. The next year Bartz visited the United States and Europe to learn about no-till planting through straw to promote infiltration and reduce erosion. He also learned that agronomist Rolf Derpsch was conducting field trials in Brazil. Eager to demonstrate that no-till technology could work in Brazil, Derpsch sent a seeder to plant a half-hectare plot of wheat on Bartz's farm. The no-till crop was a little greener and looked better than the rest of the crop. After Bartz ordered his own no-till planter, he and Derpsch kept experimenting.

Over the next several decades, South American scientists and farmers put together the system of farming we know today as conservation agriculture. Its popularity really started to grow in the 1990s, and today no-till is approaching 100 percent adoption in Argentina and southern Brazil. Because of this, the problems of serious erosion and degraded soils have been vastly alleviated across South America.

However, less than half of no-till farmers in South America use all three principles of conservation agriculture. Some leave crop stubble on their fields, others use it for cooking fuel or sell it as feedstock for biofuel production. And in response to government policies favoring commodity production, many no-till farmers in South America opt to continuously cultivate soybeans and forgo crop rotations.

When I was born, in the early 1960s, few American farmers used no-till methods. Back then more than 75,000 plows were sold each year. Yet slowly but surely, America's reliance on the plow began to wane. By 1990 fewer than 3,000 were sold annually, and in 1991 less than half that many were sold. What drove this decline?

The development of new farm equipment that could manage crop residue and plant through a surficial layer of organic matter certainly facilitated forsaking the plow. And rising fuel prices in the 1970s further spurred interest. But easy chemical weed control from Monsanto's development of Roundup (glyphosate) and genetically modified glyphosate-tolerant crops in the 1990s accelerated the adoption of no-till. And this, in turn, helped open the door for conservation agriculture as farmers began gradually adopting the other two principles as well.

Globally, conservation agriculture was practiced on less than 3 million hectares in the early 1970s. By the early 1980s, it had more than doubled to over 6 million hectares, and by 2003 it increased another twelvefold to 72 million hectares. By 2013, it had doubled again, to 157 million hectares. And yet, despite this rapid pace of adoption, only about 11 percent of global cropland is under conservation agriculture. More than three-quarters of the farmland under conservation agriculture is in the Americas, with almost half (42 percent) in South America and about a third (34 percent) in the United States and Canada. In the United States, some 35.6 million hectares of land—21 percent of the nation's cropland—was farmed using conservation agriculture in 2013. But conservation agriculture accounts for just a few percent of cropland in Europe, Asia, and Africa. In other words, there is still a lot of room for adoption.

As Lal and I talked in that Swedish hotel lobby, we kept coming back to the topic of conventional versus no-till farming—and the inadequacy with which these methods are studied, compared, and ultimately recommended. One such recent study questioned the ability of no-till practices to increase soil organic matter; another concluded that crop yields under conservation agriculture were lower than under conventional practices. Lal thought that these studies involved a fruit salad comparison—apples versus oranges versus bananas—as many of the cited studies did not leave crop stubble on the fields, and thus didn't adopt full conservation agriculture methods. It was just like the early comparisons in Africa.

A 2014 paper published in the journal *Nature* reflected some of this skepticism in conservation agriculture. This meta-analysis of 610 previous studies compared conventional practices to no-till practices, in various combinations with the other principles of conservation agriculture (cover cropping and crop rotation). Averaged across all the data, no-till practices decreased crop yields by almost 6 percent. In drylands, however, adoption of all three conservation agriculture principles increased yields by up to 10 percent over conventional practices. And after three years of no-till, crop yields from fields that followed all three principles (no-till, residue retention, and crop rotations) were indistinguishable from conventional fields. In other words, after an initial several-year transition period, there was no yield penalty for adopting conservation agriculture practices. Yet the authors of the paper emphasized, and media coverage trumpeted, that no-till farming reduced crop yields.

This, of course, riled Lal. The real test, he said, was to measure the performance of conventional farms against those that adopted *all three* principles of conservation agriculture. As it was, the wide range of practices described under the umbrella of "no-till" painted a misleading picture of conservation agriculture.

Such concerns led Rolf Derpsch and a team of leading researchers to argue for the standardization of no-till research methods to avoid just such confusion. They noted that crop yields increased where farmers experienced in conservation agriculture practices guided the conversion process, whereas yields decreased in inexperienced hands. They pointedly recommended that scientists "master the no-till or conservation agriculture system" before "attempting to research the system."[14] In their view, it seems, academics with little experience in the full system of conservation agriculture had drawn inaccurate conclusions from studies with ill-defined variables. Like Lal, they were concerned that studies reporting lower yields under conservation agriculture were making flawed comparisons.

At his talk in Malmö, Lal said our agricultural problems—and their solutions—are rooted in the way we manage soil and its world of living organisms. He showed a diagram with overlapping circles to illustrate the processes, factors, and causes of soil degradation. Erosion, salinization, and nutrient depletion were the primary processes, and the factors that drive these processes mostly lie beyond an individual farmer's control—climate, topography, socioeconomic forces, and cultural issues. But the *causes* of soil degradation were the particular practices of deforestation, plowing, and irrigation. These were the things that farmers had the power to change.

He went on to describe soils as ecological systems with thresholds or tipping points, which, after being reached, trigger significant change. One of the most important of these, he said, is when soil organic matter drops below 1 percent. Many tropical soils are already degraded to less than half this level, due to extractive farming with indiscriminant plowing and removal of crop residues. Time and again, Lal has seen this combination produce a spiral of soil degradation, human desperation, and social unrest.

Conversely, conservation agriculture could restore degraded soil and reverse the spiral to restore social and political stability. Agriculture could become a solution instead of a problem. Lal argued that while making adaptations for every site is specific and exacting, the principles of sustainable soil management are universal and simple. Putting these principles into practice requires a systems approach—just what conservation agriculture provides.

For me, Lal's take-home message was that, when it comes to farming, practices as well as the condition and quality of the soil are as important as agricultural technology.

PERENNIAL CHANGE

A couple of weeks after I talked with Lal, I caught up with charismatic septuagenarian Wes Jackson at a conference on the future of civilization at Pomona College in Claremont, California. I'd known of Jackson's work since reading his visionary 1980 book *New Roots for Agriculture* as an undergraduate. In it, he laid out another strategy for ditching the plow—growing crops that don't need to be planted each year. Ever since, he and his colleagues at the Land Institute in Salina, Kansas, have worked on breeding perennial crops.

Jackson talks like a Methodist preacher, mixing biblical metaphors, personal stories, and home-style humor, generally at his own expense. He paints agriculture as a dramatic tragedy, one in which the way we grow our food undercuts our ability to keep feeding everyone. He may look like a grandfatherly Kansas farmer, but if you get him talking about the need to save agriculture from itself, he glows with the fire of a young radical.

For forty years now, his team has worked to domesticate wild perennials and cross annual grain crops with perennial relatives in the hope of replacing conventional annual crops with deep-rooting perennial crops. In other words, Jackson's team creates GMOs, only they do it the traditional way, through plant breeding. The advantages of developing perennial crops from annual crops is huge. Not plowing to plant each year would be the ultimate in no-till agriculture, since the best way to stop using the plow is to grow crops in fields that never need plowing.

Why has he invested decades in the painstakingly incremental, old-school genetic modifications of plant breeding? A perennial ground cover is nature's recipe for reducing erosion and building soil organic matter, and thus soil fertility, over time. Developing perennial crops would not only eliminate soil erosion; it would greatly reduce the need for agrochemical fertilizers and fossil fuels.

Born on a farm near Topeka, Kansas, Jackson considers himself "objective in just the right way." After establishing the environmental studies program at California State University in Sacramento in 1971, he came to see neglect of basic principles of ecology as the Achilles heel of agronomy. Forty years old, with tenure, he returned to Kansas and started the Land Institute in 1976 to develop farming methods to build and protect soil. Appalled by erosion reminiscent of the 1930s, he began the pursuit of natural systems agriculture—the use of ecological principles to emulate natural productivity in agricultural production. On the Great Plains, that meant looking to undisturbed prairie. It soon became clear that the secret to an unplowed prairie's productivity lay in a mix of warm- and cool-season grasses, legumes, and members of the sunflower family.

Jackson realized that the prairie had remained productive since the Ice Age by keeping the ground clothed in greenery year-round, denying the bullying efforts of wind or rain. Subject to repeated fires and buffalo grazing, the prairie relied on rootstock tucked away belowground for its continuous regrowth. Reminders of how far we've gone astray since that time were on display all around the new institute, in neighbors' fields that bled brown runoff in spring storms.

But how to transform this land? Jackson wanted an herbaceous polyculture founded on perennial grains instead of annuals. Yet all of humanity's major grains were annual grasses. If he wanted perennial grains, he'd have to make them.

So the Land Institute set about trying to transform wild perennial wheatgrass into a perennial grain, their goal to build a domesticated prairie that would produce grain. Jackson predicted it would take fifty to one hundred years to do the research and breed the plants. As it turns out, his team is ahead of schedule—way ahead.

That's a good thing. Time is not on our side. Lately, Illinois corn yields have been dropping by a bushel per acre a year. Modern fertilizers have proven inefficient, as grains take up just 40 to 70 percent of the nitrogen farmers apply to fields. Producing nitrogen fertilizer through the Haber-Bosch process is both energy-intensive and requires a catalyst, a temperature of 400°C, and 350 times atmospheric pressure. Global pesticide production doubled after Rachel Carson published *Silent Spring* and then doubled again in subsequent decades. Modern agriculture may not be terribly efficient, but it has proven quite effective at degrading the soil and the life it supports.

Right from the start, experts thought that Jackson's idea that perennials could produce high yields simply would not work. But he saw deep roots as the best investment plants could make. So his team planted thousands of seeds each year, looking for plants that showed promise as grain crops. Then the breeding began. Doing this over and over, they produced a deeply rooted perennial grain that shows tremendous promise.

THE NEW PERENNIAL

With practiced flair, Jackson unrolled a large scroll across the conference table. People around the room gasped as a life-size photo revealed a side-by-side comparison of the root systems of conventional wheat and Kernza, the new wheatgrass variety. The three-foot-long, stringy root system of the wheat plant looked anemic next to the Kernza's beefy, ten-foot-long dreadlocked root mass. Behold, he proclaimed, the first perennial grain in history—and humanity's first new major crop in over 3,500 years. This achievement showed that wild perennials *can* become grain crops, an exciting possibility for farms across America, and potentially around the world.

Nearly forty years after his mission started, a thrilled Jackson showed off a video of the first mechanized harvest of a field of Kernza. The new grain had already been used to make beer, bread, and whiskey. Now Kernza was being harvested commercially, though on a limited scale. He seemed positively giddy.

As well he should. With more breeding, perennials will one day generally outyield annuals, given their longer growing season. Perennial crops also get a head start on developing a canopy cover and shading out weeds. Initially, Jackson worried that Kernza's success would cause it to be planted as a monoculture, but his crew has been making progress on other perennial crops. They now have perennial sorghum undergoing field trials in Central Africa. Perennial sunflowers and chickpeas are in the works too. The Land Institute is also experimenting with coplanting legumes with Kernza and allowing cattle to graze on Kernza stubble. There is still much left to do, of course, but Jackson sees his dream of a domesticated prairie as a polyculture with multiple interplanted crops becoming an agronomic reality.

As I was writing this chapter, I got an email announcing that Jackson was stepping down as president of the Land Institute, just in time for his eightieth birthday. Not everyone gets to see big dreams bear fruit in their lifetime; Jackson is one of the lucky ones. He now is confident that, through applied ecology, we can harvest soil health as a consequence of agricultural production. This is revolutionary—the seeds of truly transformative change.

While Jackson's vision may be a sure way to solve the problem of soil erosion and sustain agriculture over the long run, plant breeding is a slow game. Even Jackson estimates it will take decades to finish his work. And there is no commercial support for development of perennial crops. No seed company will touch perennial crops, for obvious reasons: What kind of business model is it to only sell a customer seeds once? Seed companies, like drug companies, want repeat customers year after year. And this is what annual seeds guarantee those who sell them, especially proprietary and patented ones. So what can we do in the meantime?

In theory, conservation agriculture practices could be adopted tomorrow—without delay. But could they really work on farms large and small in both the developed and developing worlds? Few studies have examined the effect of adopting *all three* principles of conservation agriculture. So to find out, I decided to go and see for myself and embarked on a six-month tour of farms doing this on several continents. The first stop on this journey busted another myth of modern agriculture: the contention that no-till farmers *need* to use a lot of herbicides and fertilizers.

6

GREEN MANURE

There is no profession which for its successful practice requires a larger extent of knowledge than agriculture, and none in which the actual ignorance is greater.

—Justus von Liebig

I first met Dwayne Beck, director of Dakota Lakes Research Farm, at the 2014 World Congress on Conservation Agriculture. From the get-go, his presentation had me riveted as he described how farmers around Pierre, South Dakota, transformed a landscape once famous for generating epic dust storms into highly productive farms with healthy soil.

So on a sunny day the following April, I flew into Denver and made my way to the farthest end of the airport, down a long hallway where a small plane waited at the last gate in the terminal. During the hour-and-a-half flight, I looked out at lean brown hills dotted with stock ponds and gullied scars exposing flat-lying geology. Solitary farms stood out as patchy green circles and rectangles scattered across the terrain. As we circled to land, the first stop on my six-month tour, I saw the dam-drowned Dakota lakes along the Missouri River, a blue ribbon of life snaking through parched land.

Beck greeted me at the airport and drove us to his modest ranch-style house, located near what was once the edge of an ice sheet. Half the town rests on glacial till, compacted debris formerly overrun by ice. The other half sits on outwash sands laid down by long-gone melt-water streams shed from the massive glacier. It's a landscape of transitions, and Beck was a driving force behind the most recent one—the widespread adoption of no-till farming and integrating cover crops into rotations.

Sitting around the table after dinner, I wasn't surprised to learn that Beck grew up on a local farm. Affable, solidly built, with huge hands and thick black-rimmed bifocals, he looked the part of a farmer, right down to the rebuilt thumb, which had been nearly taken off in a farm machinery accident. Born in Platte, South Dakota, in 1951, Beck had attended a one-room country school and was ten years old before his family had running water. When he headed off to college at Northern State University in nearby Aberdeen, he found the indoor bathrooms a welcome luxury. After getting his B.S. in 1975, he taught high school chemistry before enrolling in graduate studies and completing a Ph.D. in agronomy at South Dakota State University in 1983.

Working on a thesis analyzing the efficiency of fertilizer uptake, he noticed that farmers were losing a lot of water in runoff from their fields. When he conducted sprinkler experiments to test how much water sank into the soil, he found that plowed fields tended to crust up and produce a lot of runoff, while, much to his surprise, fields left untilled produced no runoff. In these fields, even intense rainfall would sink into the soil.

To Beck the implication was clear—better to not till than to till. After he completed graduate studies, Beck kept experimenting with no-till, eager to put an end to the clouds of dirt that closed the highways anytime the prairie winds kicked up after plowing.

In 1990, he became research manager at the newly established Dakota Lakes Research Farm. The farm itself was an experiment. Farmer-owned but run in cooperation with South Dakota State University, it had an all-farmer board of directors who'd seen the damage tillage did to their soil. They wanted alternatives.

For twenty-five years, Beck has run the farm with an eye toward developing farming systems that work better for farmers and the environment. Through his research he's found that when you restore soil biology, you don't need to plow or rely so much on chemical inputs. But this requires a new system of farming in which no-till methods and cover crops are used to grow a diversity of crops in complex rotations.

At first, the South Dakota Wheat Commission was skeptical of what no-till and crop rotations would do to the wheat harvest—and to their budget. But they found that wheat yields actually went up, as did the yields of other crops. By all accounts, it's been astoundingly successful. Beck reports that annual crop production in the region around Pierre increased by $1.6 billion compared to 1986. The whole system was more productive and *almost* everybody celebrated; the best approach environmentally proved the best approach economically except, of course, for fertilizer, herbicide, and insecticide manufacturers.

The next morning, we drove north, taking the long way to Beck's farm to visit various Dakota Lakes board members on theirs. Heading past the enormous earthen Oahe Dam, we stopped to take in vistas Lewis and Clark may have seen but wouldn't have understood in geological terms. They wouldn't have known that the river they'd laboriously paddled up marks the edge of an ancient ice sheet, where boulder-strewn fields rolling off to the glaciated east give way to gentle terrain to the west. For Lewis and Clark, this was no doubt just more land to cross as they explored the new Louisiana Purchase looking for a route to the Pacific.

As we sped across a prairie-grass sea, Beck told me how farmers worked their fields back in the 1970s. They grew wheat one year in a field and then plowed it up and let it lie fallow the next year. This made a wheat-fallow rotation that was good, if you didn't mind losing soil organic matter due to regular tillage. Farmers eventually became hooked on fertilizers as this pattern burned through soil fertility.

After we'd stopped by a few farms, I started to recognize some common themes. Becks' board members—Marv Schumacher, a big, quiet guy, Kent Kinkler who kept a rack of guns with bayonets in his office, Mike and Ann Arnoldy, a brother and sister team with a sweet black lab, and the charming couple of Ralph and Betty Holzwarth—all professed a desire to restore their soil, so that their kids would inherit farms capable of supporting them when the time came for them to take over.

Back before each of these board members had met Beck, they'd all done high-disturbance, input-intensive farming. But when droughts and rising input costs pushed them to the edge of economic viability, they were ready to try something different. Soon after switching to no-till, they noticed big reductions in their water use, and substantial savings from lower diesel and chemical bills. After several years, they started to see their soil change too. Crop yields came back up to or surpassed what they'd been before, but with lower input costs. No-till farmers doubled their harvests by planting their entire farm each year instead of just half their fields, as they did under the wheat-fallow rotation. They also started to think differently about their soil—and the life within it. When I asked Kent Kinkler why he didn't have any cows, he replied, "It's not that I don't have any livestock, it's that mine are microscopic." These South Dakota farmers had shifted not only the way they farmed but the way they thought of their ground, from chemical reservoir to a sea of biology.

All this translated into a better bottom line. The best wine cellar I've ever seen, stacked floor-to-ceiling with fine French reds, belonged to one of these South Dakota farmers. Beck told me how he once took a group of economists from Kansas, who insisted that no-till farming couldn't make money, around to see farms in this area. He didn't have to say a whole lot. The no-till farms had new grain silos, new equipment, and new homes. The conventional tillers didn't. The economists got the point.

SHOW, DON'T TELL

How did Beck convince farmers across the region to adopt a whole new system of farming? For starters, he's a rare breed, a Ph.D. researcher who can drive farm machinery and design new parts for seeders, then build them and try them out. He knew that showing farmers a Power-Point presentation of his research results wasn't going to cut it. Instead, he started by convincing a few guys to give his ideas a try and donating the use of some of his equipment designed for no-till to the cause. Then, when other farmers witnessed their neighbors' success, they followed suit. Beck knew that farmers prefer to get their information from other farmers. It helped when the incomes of his early adopters increased. South Dakota went from almost all tillage in 1990 to more than three-quarters no-till by 2013. This radical shift transformed farming practices over a couple of decades.

Driving back into town, it became clear that folks here trust his advice. The cell phone in his breast pocket rang constantly with farmers asking about what to plant, what kind of equipment to use for this or that task, and what the value of chicken manure is as fertilizer. I wasn't surprised. He is one of them, well cast in blue jeans and a knit shirt, chewing on a neon-orange golf tee.

Beck planned this whirlwind farm tour to make an impression. And it did. On the several-hundred-mile drive around South Dakota, I didn't see much bare earth. As we crossed hills that reminded me of California in summer, he lamented, "I've never seen it so dry this time of year." No rain to speak of had fallen since August. Before no-till this would have been a disaster. He said it was drier than it had been in the 1930s. Yet there was hardly any dust, and no wind erosion—thanks to the widespread adoption of no-till.

But the soil wasn't buttoned down everywhere. Across the river from Beck's farm, we passed a tractor spewing great clouds of black earth into the sky. Appalled, I turned to Beck and he told me that this is still common among some tenant farmers with no apparent interest in the long-term health of the land.

After crossing the river to a flat terrace along the north bank, we arrived at Dakota Lakes Research Farm, 800 acres of land just south of two-lane Highway 34. The entrance, a gravel road on the west side of the farm, runs right along the famed 100th meridian. This geographic milestone represents the western edge of the Great Plains, the border between rain-fed and irrigated farming. About a quarter of the farm is irrigated, and on that day a towering four-span sprinkler slowly "walked" across the fields like a giant dripping robot. The terraced surface drops down to the river through a gully system littered with rusting farm machinery.

On their way up the Missouri River, Lewis and Clark had crossed this land. Two centuries later, Beck used their journals to establish that the native vegetation was tall-grass prairie and mixed-grass prairie. He then planted a fringe of prairie grass in a ring around the farm to provide habitat for pollinators and the native predators that help reduce pests in the fields.

Beck wanted to show off the farm's new zero-net-energy building, so we drove over to a big gray shed with solar thermal panels and photovoltaic cells on the roof. My house could've fit inside the shed and sprayed-on, soybean-based foam insulation covered its girders and walls. This may look funky, but it works—though it was seriously hot and dry outside, it was nice and cool inside. Beck assured me the opposite is true in winter.

Canola or soybean oil made on site heated the half of the building with offices. Against the north wall, a giant screw press squeezed oil from soybeans, spitting out soy-cake strips that looked like giant onion rings. This "waste" is mostly protein, used to feed animals who, in turn, produce manure to return to the fields. This, Beck explained, is energy-efficient nutrient recycling. His ultimate goal is for Dakota Lakes to become fossil fuel–neutral without losing nutrients and organic matter from the soil, or native predators from the land. He's certain that doing so requires transformational, rather than incremental, change.

Understandably, farmers do not generally take drastic measures before seeing evidence of success. They are not impressed with a few little test plots. They want to see new ideas working at full scale—on a real farm—before they try them on theirs. But Beck worries that most agricultural researchers and extension agents who advise farmers are specialists well versed in particular practices. They generally focus on incremental steps, like trying a different herbicide, instead of transformational change, like figuring out how to use natural cycles to minimize the need to use any. So it's no surprise that there are few no-tillers in parts of the state where agency research farms still till. If the farmers don't see researchers trying something new, then why would they?

Of course, there's more to it than a lack of demonstration models. Beck worries that few university faculty and rookie-Ph.D. county extension agents actually have experience farming. Coupled with deep cuts to state and federal agricultural extension programs, he fears we're turning the practical knowledge-keeper role over to industries with vested interests in promoting their own products. Most companies are going to be less than enthusiastic about funding research on changes in practices that don't lead to a product they can sell. Unfortunately, the USDA seems to emphasize corporate partnerships and the testing and developing of new products over systems research to find cost-effective practices that use less inputs and still meet farmers' needs.

Beck's farm prioritizes research that benefits farmers. His first loyalty is to the five hundred members who own his farm, which he sees as a learning community. Many of the key farmers involved are the children and grandchildren of the people who first started the farm. He thinks it's worked because farmers want continuity of community, and introducing no-till to South Dakota has breathed new life into theirs.

Dakota Lakes Research Farm is funded by the profits produced by the farm, funding from South Dakota State University, and member contributions for special projects. This insulates their work from special interests and politics. Due to limited manpower and limited capacity to invest in new machinery, the farm adopted crop rotations and water conservation measures and pursued agronomic choices based on ecological principles. Over time, practices at the farm changed along with new technology and knowledge about how to promote soil health and control pests and weeds using cover crops and crop rotations.

Midmorning, a soil physics class from South Dakota State arrived for a tour, the fourth group this week. Twenty students and their professor piled into the shed. Once they were settled, Beck started off with a big-picture goal: he aims to make Dakota Lakes carbon-neutral by 2025. Why? Because 80 percent of what wheat farmers spend on operating costs can be traced to fossil fuels. One hundred years ago it was zero, and one hundred years from now it'll be zero again. We'll just have to learn how to farm the old-fashioned way with modern technology, without sacrificing yields or the soil itself.

I was still the only one taking notes as Beck herded us outside to his mobile classroom, a flatbed trailer with bleachers mounted four rows high and hooked up to a green John Deere tractor. We piled on, and the first thing that Beck pointed out was the native switchgrass fringing the farm.

In our bleacher buggy we lumbered down the well-traveled dirt road that bisects the farm. Each test plot we passed had been thoughtfully overseen by Beck. A few have grown corn continuously for years. Others receive various combinations of crops, rotations, herbicides, and fertilizers.

But the next stop on the tour was all about water. Beck told the class about how he'd gotten into no-till from studying irrigation runoff. He'd seen people putting pivot sprinklers onto plowed soils, which promptly formed a heavy crust that wouldn't let water sink into the ground. This, of course, defeated the whole point of irrigating. How does no-till help? It preserves the capacity of the ground to absorb water instead of shedding it as erosive runoff.

And no-till helps in other ways too. Leaving corn stubble standing creates low-velocity airflow at the ground surface, which reduces evaporation. Beck spelled out the net effect for the students: "You stop doing tillage, you save moisture."

Compaction matters too, Beck went on. Running heavy machinery across fields compacts the soil, crushing the void spaces that hold water crops can use. That's why everything on the Dakota Lakes farm has big fat tires, inflated to just 11 psi, and only 7 psi if the ground is wet. That's about the pressure one generates standing on one foot.

No-till also increases the organic-matter content of the soil, which affects its water-holding capacity, something vitally important in the semiarid Dakotas, where summer soil moisture is a make-or-break issue for crop yields. An increase of organic matter content from 1 percent to 3 percent can double the soil's water-holding capacity, while helping to prevent water logging that leads to the anaerobic conditions that favor soil-dwelling pathogens. "They told you that you can replace organic matter with fertilizer," Beck told the students. "They lied."

Meanwhile, the irrigated no-till fields on the farm used just half the water of a similarly sized plowed field. This, Beck concluded, is climate resilience. To illustrate his point, we stopped at a pair of neighboring plots that had been irrigated with an inch and a quarter of water the previous night. One of the plots had bare soil, with no crop residue left on the ground surface after the previous year's harvest. The ground was cracked, with weeds shooting up through a thick crust and signs of erosive runoff over the surface. The neighboring plot had residue from last year's crop—and no cracking, no crusting, and no signs of runoff. I was impressed that a single irrigation session could produce such an obvious difference. As students dutifully checked their cell phones, I closed my notebook and let my right hand rest for a few minutes.

NO ROOM FOR WEEDS

When he was starting out, just about everyone told Beck that no-till methods invited weeds and crop diseases. But he came to realize that "if tillage was good at eliminating weeds, all the weeds in the U.S. and Canada would be gone by now." In practice, he discovered that the less he disturbed the ground, the fewer the weeds he had to contend with.

Over time, Beck came to see that the best weed control was competition from the canopy of a well-nourished crop. Effective weed management, he realized, is not about killing weeds but about taking away their opportunities. Leaving a thick residue from the prior crop makes it hard for weeds to get going, and no-till planting gives the crop a head start, thereby depriving weeds of water, space, and light. Beck finds he can use cover crops in crop rotations to out-compete weeds and reduce herbicide use. Cover crops also provide side benefits, such as adding carbon and nitrogen to the soil, which reduces the need for fertilizers. It's the agricultural equivalent of preventative medicine. If you manage a crop rotation right, your fields will be healthier in general and weeds never get out of control.

Back in the early 1990s, Beck was just starting to talk about using crop rotation to control weeds when corn tolerant to the herbicide family called ALS inhibitors first came out.[15] At one conference, a young farmer asked why he should consider crop rotation when new herbicides would take care of any weeds. Beck bluntly replied that weeds would continue to evolve and eventually develop resistance to every new herbicide they meet. The next thing he knew, he—and the president of his university—received a letter from the company that made the herbicide, asking for a public retraction. Beck wrote back that he would be willing to retract his prediction if, in three-years' time, weeds at Dakota Lakes didn't develop resistance. He then set up a field trial on which he spread seeds from a neighboring field where ALS inhibitors had been used for years and started applying herbicides at twice the recommended rate. In only four months, he was growing resistant weeds. He never had to write that letter of apology.

This story is now playing out on a larger stage. For years, Monsanto dismissed concerns that weeds would develop resistance to glyphosate (Roundup). Yet between 1996 (when glyphosate resistance was first noticed) and 2011, at least nineteen weed species had developed glyphosate resistance. At that time, a review of herbicide-resistant weed management, led by an industry scientist, argued that "it is of paramount importance that other weed management alternatives be identified and implemented quickly" because of the growing problem of herbicide-resistant weeds.[16] Shortly thereafter, regulatory wrangling began over an herbicide designed for use with corn and soybean seeds engineered to tolerate *both* glyphosate and 2,4-D. Beck thinks that this move will inevitably breed weeds resistant to both herbicides. He prefers a different solution: outcompete weeds with other plants and keep herbicides effective for when farmers really need them.

Farmers generally love the built-in weed control of corn genetically engineered to survive glyphosate while the surrounding weeds die off. But there's a down side, Beck told the class, pointing out the herbicide-resistant weeds along his fence line, creeping in from the neighboring farm—at some point the weeds will develop resistance. Dakota Lakes grows genetically modified corn and soybeans, in part because its hard to find non-GMO corn and soybean seed in the United States. But they use only a fraction of the glyphosate they formerly did because they just don't need much anymore. There are shockingly few weeds in Beck's fields—I had a hard time finding *any*. That's the value of a continuous ground cover.

Beck is not alone in finding that cover crops and crop rotation can combat weeds. One USDA research agronomist in particular, Randy Anderson, led studies of weed pressure on crops under different tillage systems in Colorado and South Dakota and found three- to twelve-fold reductions in weed density between two-year (corn-soybean) and four-year (wheat-corn-soybean-pea) rotations. Changing from the region's conventional wheat-fallow to a two-year, no-till rotation with both cool- and warm-season crops lowered weed density enough to cut herbicide inputs in half. Implementing a four-year rotation eliminated the need of herbicides for some crops altogether. The take-home message is that cover crops and rotations that leave more crop residue on the ground surface provide greater weed suppression, up to more than 80 percent compared to bare soil.

GROW-IT-YOURSELF FERTILIZER

Cover crops not only greatly reduce the need for herbicide, they also help reduce fertilizer use. If, that is, you plant the right crops: those that fix nitrogen and build up carbon in the soil. Beck thinks about farming as feeding the soil so that the soil can feed the crop. He says that in a no-till corn-soybean system, corn and soybeans both benefit from exchanging carbon and nitrogen with microbial life in the soil. But when tillage disrupts the microbial connections, corn and soy become competitors.

Beck considers crops a "catch and release" system for nutrients. Other researchers have also found that growing nitrogen-fixing legumes as cover crops can partially to fully offset the need for nitrogen fertilizers, and increase crop yields. Cover crops help reverse soil degradation by taking up nutrients from the soil and concentrating them in organic matter to be released in plant-available forms upon decay. One such study, a 2002 report of long-term farming practices in the western Corn Belt found that crop rotations containing both legumes and non-legumes greatly increased soil nitrogen availability over long-term monocultures, which decreased soil nitrogen.

Some rotations produce more residue and put more carbon in the ground than others. Beck is currently experimenting with growing two crops at once in the same field. Instead of planting on two thirty-inch rows, they'll use three twenty-inch rows, with two rows of corn and one row of alfalfa. The twenty-inch rows of corn will produce the same crop density as typical thirty-inch rows of corn, with the help of the row of alfalfa, which will feed nitrogen to the corn. At least that's the plan.

However that experiment turns out, the effects of tilling on organic matter and nitrogen content in the soil are well known. "Send Grandpa the bill," Beck told the students. "He used up all that nitrogen and organic matter."

A strong breeze came up, and I grabbed for my hat as Beck started to explain how he and his colleagues could use wind to make nitrogen fertilizers, right there on the farm. Windmills can provide enough power to split hydrogen from water. Passing the hydrogen along with nitrogen gas (the dominant component of air) through a high-temperature, pressurized reactor with a catalyst creates ammonia gas (NH_3) that, upon cooling, can be used as a fertilizer. Once in place, wind is all that such a system needs to produce fertilizer for the next season. And if there's one thing the plains are blessed with, it's wind.

In fact, new wind farms are sprouting up all across the region. A 2015 University of Minnesota study showed that such an approach could provide for cost-effective, community-scale ammonia production using wind power and greatly reduce agricultural greenhouse gas emissions. An even more novel approach would be to use pond-cultured cyanobacteria to produce ammonia. Researchers have proposed that this could replace the Haber-Bosch process and that, if widely adopted, would be a much more environmentally benign method of fertilizer production.

But we can also be smarter about how we apply fertilizers. That's the idea behind precision fertilization. Systems designed for no-till planters, like the ones Guy Swanson sells, can place a little fertilizer just a couple of inches off to the side of the seed, giving the young plants what they need to grow. It also allows farmers to use much less fertilizer and reduces the amount that runs off fields to pollute waterways. While Beck thinks we eventually might be able to eliminate nitrogen fertilizers altogether, a little bit of nitrogen precision-applied in appropriate proximity to seeds during planting produces a 10 percent boost in yield. Still, he sees precision agriculture as just one of many tools needed in a modern farmer's toolkit.

Then there's phosphorous, another nutrient critical for plant growth. Beck likes to say that a 120-car train of soybeans heading to market carries off about half a million pounds of phosphorus. Most phosphorous in soils is not soluble and doesn't show up in soil tests, since it's locked up in minerals, stable oxides, and organic matter in the soil. But there is a way to get at this phosphorous and make it available to plants.

Mycorrhizal fungi, and certain bacteria, solubilize phosphorus *and* deliver it to plants. Why? In exchange for access to the sugary exudates the plants push out their roots. In this way, fungal hyphae serve as root extensions for plants.

Unfortunately, tillage breaks the hyphae strands, chopping them up and destroying their connection to plant roots. This means that if you till, you have to add phosphorus because the mycorrhizal fungi aren't there in sufficient numbers to do their job. But with enough of the right mycorrhizal fungi in the soil, crops can get the phosphorous they need at much lower soil test levels—so much so that applying a lot of fertilizer doesn't boost yield. This means that farmers face a choice: till and then fertilize a lot, or skip tillage and fertilize less, if at all.

Beck stopped the bleacher buggy and, as we spilled out onto a stubbly field, he called our attention to the corn stalks. He bent down and scraped away the residue on the ground to expose a wormhole, then stuck his pen into it as far as it could reach. The holes go down four feet, he said, and there's one about every nine inches. Worms that live in permanent burrows drag crop residue down into their lairs, digesting it into castings rich in soluble nutrients. All those deep holes also allow water to sink into the ground, where thirsty plants can later access it. Worms are like tiny livestock, eating organic matter and fertilizing a farmer's field. We need them there, wellfed and happy. Plowing is like setting off a bomb in their living room, first destroying their homes and then, when the bare soil at the surface has turned into a water-resistant crust, drying up their water well.

As Beck walked us around the farm, he dug holes to show the effects of different practices on the soil. His long-term, no-till soil was dark brown, crumbly, and moist just below the surface—despite this being the driest year in memory. It's a major improvement from the dusty khaki earth he started with.

PEST SELF-MANAGEMENT

Beck likes to tell visitors how he used to teach a class in which half the students usually came from farm families. He'd survey them about their parents' pest problems and compile a list on the board. Then he'd go down the list and tell the students what style of rotation and disturbance regime (plow versus no-till) their parents used based on the pests they had—and got it right nearly every time. The students were always flabbergasted.

Beck says that it's a myth that we need to regularly employ insecticides and herbicides. There are other ways to manage pests, and lessons from South Dakota apply in other regions too. Most farms in the Corn Belt (Iowa, Illinois, Nebraska, and Minnesota) practice a simple corn-soybean rotation. At any one time, more than 98 percent of Iowa farmland is covered with corn or soy. With such regularity and uniformity, we might as well be ringing a dinner bell for pests and crop diseases. And then when they come running, we pull out the pesticides.

Much like modern-day antibiotics, broad-spectrum pesticides take out the good with the bad. The seeds of many crops are now routinely treated with neonicotinoid insecticides, systemic toxins that migrate into the sap and spread throughout a plant. This not only kills insect herbivores but also the insect predators that help keep the pests in check.

For example, a 2015 study in the *Journal of Applied Ecology* reported that soybean seeds treated with a widely used neonicotinoid (thiamethoxam) impaired or killed the predatory beetles that eat slugs. The slugs, freed from the threat of predation, proliferated and reduced soybean yields by 5 percent. So although the seed treatments were intended to increase yields by getting rid of pests, the resulting disruption of biological pest control proved counterproductive.

A global team of scientists reviewed more than eight hundred peer-reviewed papers published over the past several decades and concluded that the agronomic benefits of neonicotinoids remain unclear. Few studies demonstrated a net benefit on crop yields, while some showed net economic loss. In addition, the team found strong evidence for serious adverse effects on beneficial nontarget organisms, such as insect pollinators like butterflies and bees; ingestion of even a few treated seeds can prove lethal to small birds.

The corn rootworm beetle became established as a perennial problem only after farmers in the Corn Belt and on the irrigated plains began growing corn on the same land year after year. Rootworm beetles would eat the corn silks and lay eggs at the base of the plant. The following spring, the eggs would hatch and, if there were corn plants on the field, the larvae would chow down. Soon, land continuously planted in corn required large amounts of pesticides to prevent heavy losses to these ravenous beetles.

But when soybeans were included in the rotation, there was no corn root feast awaiting the newly hatched corn rootworm larva. This helped, but it didn't completely solve the problem, since a fraction of the eggs hatched the *second* spring after being laid. The simple corn-soy rotation ended up favoring the survival of beetles with this two-year life cycle because the interval between successive corn crops was held constant. Soon the beetle population began to follow this pattern. On top of that, some female beetles lay their eggs on soybean fields. In a predictable corn-soy rotation, their eggs would hatch into a cornfield. Whichever way the beetles adapted to a simple corn-soybean rotation, the problem started all over again.

The key to pest control, Beck suggested, is to understand the habits of pests. Consider the case of the western bean cutworm and corn earworm. A mother earworm feeds on pollen and lays her eggs on an ear of corn. Only one of her darling cannibalistic babies survives—the one that eats all the others. When Pioneer, a subsidiary of DuPont, developed Bt corn that killed the corn earworm, another problem arose. Before Bt corn, the western bean cutworm wasn't much of a problem, since the ravenous corn earworms not only ate their siblings but all the bean cutworm babies as well. But when Bt corn took out the earworm, it created a golden opportunity for the bean cutworms. All of the bean cutworm larva survive, causing more damage to the ear. The new technology to control one pest created a new, even more problematic one. How is that progress?

Beck offered the students another example from an earlier experiment conducted at Dakota Lakes. U.S. Department of Agriculture and South Dakota State University researchers introduced a thousand corn rootworm eggs per foot of crop row to a continuous corn plot that had been neither tilled nor received any broadcast insecticide since 1990. When the pests didn't cause any noticeable problems, they dug up cores and found the reason: the soil had more than a billion predatory insects per acre. "All you'd need to screw this up," he said, "is to broadcast-spray insecticides." Kill off the predators and the prey population explodes—just like what happened to deer in North America after we killed off cougars and wolves.

So when Beck discovers a pest problem, his first step is to determine what provided the opportunity to the pest—just like for weed management. Then, figure out what to do to mitigate it without whacking both good and bad insects with chemicals—and creating new opportunities for other pests. Complex rotations with irregular intervals can keep pests from adapting. Doing this, he hasn't needed pesticides for over a decade.

By changing from a simple corn-soybean rotation to a more complex corn-corn-soybean-wheat-soybean rotation, Beck increased his soybean yields by 25 percent, from 63 bushels an acre to 79 bushels an acre. At the same time, corn yields increased as well, rising from 203 bushels an acre for continuous corn to 217 bushels an acre for a corn-soybean rotation, and 235 bushels an acre for the more complex rotation. The whole system became more productive under a diversified rotation. And because it uses fewer inputs—less diesel, fertilizer, and herbicide—it's even more profitable.

HIGH-TECH NO-TILL

So why don't more farmers adopt this system of cover crops and complex rotations? A big reason is that subsidized government programs, like crop insurance, penalize farmers who use complex rotations, and sometimes don't even allow them. Beck believes that taking away artificial subsidies and changing crop insurance policies would shift more farmers toward adopting practices that can maintain yields while reducing both input costs and environmental damage.

That afternoon, after the students left, we drove north to Gettysburg to visit Cronin Farms, a large family farm that adopted Beck's approach a decade ago. Along the way, Beck related how back in the 1980s he had to drive through blowing dust, praying that he'd make it to the Sully-Potter county line. In the 1960s and '70s, Sully County had an influx of farmers from Oklahoma and Texas, where tillage-based wheat-fallow systems were the norm. In contrast, those coming to Potter County were predominantly European immigrants accustomed to more diverse, integrated livestock and grain operations. The diverse farmers more quickly saw the advantages and adopted no-till. Back in those days Beck knew that the skies would clear if he could make it to Potter County.

Passing through the town of Onida, I was impressed by the towering grain bins, rising high above the prairie. A row of eight shiny new ones sat beside old concrete ones, testifying to the recent increase in harvests and prosperity. The town had almost disappeared in the early 2000s, but it's coming back thanks to no-till. The price of land reflects this new productivity: between 1990 and 2015 the price of an acre shot up tenfold, from around $300 to $3,000.

As we pulled in to Cronin Farms it seemed to stretch from horizon to horizon, 9,000 cropped acres and 11,000 acres of native prairie pasture. A compound of farm buildings, corrals, and grain silos stand together like a cultural island amid a prairie sea.

The Cronin brothers, Mike and Monty, own the farm. But it was Dan Forgey, the farm manager, who greeted us. Clean-shaven, with thick black angular glasses, jeans, and gray hair tucked under a John Deere hat, Forgey is earnest and solid, the kind of person you trust from the get-go. He's worked at Cronin Farm for 45 years of its 104-year history.

When Forgey first started at the farm, it was a conventional operation. He says that he spent the first half of his life destroying the soil before seeing the benefits of no-till. Now they use thirteen different rotations on the farm and switch things up year-to-year, depending on the weather. They've been listening to Beck—and it's working.

Forgey told us that he likes how crop rotation spreads the workload more evenly throughout the year, and that the Cronin family is committed to "doing things right" on the farm. "We're in the middle of a big change—we have to get it right this time." A decade ago he, along with everybody else, sprayed to control invasive cheatgrass. Now he manages it with crop rotation.

Along with cover crops and crop rotation, they've embraced high-tech, precision agriculture. It helps them track rotations and yields, and to customize (and minimize) their fertilizer application rate. They place all their nitrogen belowground at planting, two to three inches off to the side of the seeds, instead of spraying it all over the ground surface. A little phosphorus goes right into the trench along with the seeds. This saves them 15 percent on fertilizer outlays, and in trips across the field. They've been hitting yield goals of 130 to 135 bushels per acre for dryland corn and 65 to 90 bushels per acre for winter wheat, harvests comparable to county averages but with much lower input costs.

About twelve years ago, Forgey learned about the effects of fungicides on other friendly soil life the hard way. After spraying fungicide on their spring wheat, a devastating bacterial leaf blight took hold. He thinks the fungicide knocked out the beneficial fungi that kept pathogenic bacteria in check. Now, Forgey says, they only spray if they have to. Besides, anytime they spray, there is a cost. And like farmers everywhere, Forgey doesn't like to spend money—even Monty and Mike's.

Forgey has seen how "the soil tells its own story," and how what's belowground supports what's on top. To build up mycorrhizal fungi and organic matter, they now grow an impressive variety of crops—winter wheat, spring wheat, oats, corn, sunflowers, barley lentils, field peas, flax, and alfalfa, as well as teff for grain or forage.[17] So far, they've increased their soil carbon by 1 percent. That may not sound like a lot, but Forgey says that each percent of organic matter holds about $600 worth of nutrients an acre—he's looking at his soil in a whole new way. Forgey hopes to raise their soil organic matter even more by growing high-carbon cover crops that leave more residue on the fields.

Forgey uses a tool that Beck doesn't yet—cows. The Cronin's 900 cows graze on cover-crop stubble and manure the fields. Forgey took us out to a no-till field at the edge of the farm—right across from a neighbor's high-disturbance field—where cows had grazed cover crops for December forage. The ground was still partially covered with the remains of radishes, oats, flax, and turnips. I knelt down and dug my hand into the soil, pulling up rich blackish soil, moist just below the surface. The silt crumbled away with a nudge of my thumb. Forgey used a pocket thermometer to measure the soil temperature; though the air temperature was 80°F, it was 55°F under the crop residue.

Forgey then led us across the dirt road separating his field from his neighbor's. There wasn't much crop residue on the bare ground where the neighbor direct-seeded using a hoe drill, a practice that Forgey called "aggressive, high-disturbance no-till." Beck chimed in to say that this style of planting disturbs too much of the soil surface and soil life. The difference between the two no-till fields was like night and day. Here, on the neighbor's field, the soil was much lighter in color, and I could neither dig a hole with my fingers nor crumble blocks of the platy soil we dug up with a shovel. I couldn't even push my thumb into the brick-like blocks of dirt. The ground temperature was 74°F, and it was bone-dry below the surface. The soil-moisture difference alone between these two adjacent fields drove home Rattan Lal's comment that not all no-till systems are equal.

When the wind picked up enough to knock our hats off, I could see how this fine silt soil would readily blow away if plowed, dried, and exposed. This was why the Dust Bowl happened, and why crop residues are essential for keeping the soil cool, moist, and on the ground— where it belongs.

Forgey drove us down to a big circular pivot-irrigated field by the river, where Mike Cronin was planting corn. When we got there, Mike, tall and dirt-covered in his orange shirt and jeans, offered to show me around his enormous John Deere–green planter. We started at the business end, then walked along the machine to examine its guts, an odd collection of disks and wheels. Though it looks complicated, it's actually pretty simple and remarkably clever. Mike explained the process: First the smooth cutting wheel opens a narrow slot (trench) in the soil and a spot of fertilizer is applied several inches off to the side of where the seed will be placed. Two angled "residue managers" sweep excess residue from the row area without disturbing the soil. Next a double-disc seed opener cuts a trench in the undisturbed soil and places the seeds in the bottom of the trench. A small amount of fertilizer is added to the trench after a seed-pressing mechanism has pushed the seed into the soil at the bottom of the trench. Finally, a pair of spiked closing wheels that angle in from either side close the seed trench back up. After the planter has done its thing, it's hard to tell that the ground has even been disturbed.

Mike asked me if I wanted to take a ride, and I enthusiastically followed him up into the cab and closed the bubble door, feeling like I'd entered the command center of a two-seater prairie starship. On the far side a stack of monitors showed satellite-generated GPS run off an iPad and the output of John Deere sensors that track performance of the unit. Mike took the pilot's seat and loaded a memory stick with a field-specific program.

He fired up the engine. As we crawled across the field, I followed our progress on a bank of readouts. The GPS system kept track of our location and, in the wake of our passage, the split-screen display showed the downward force exerted by the tires and the application rates for fertilizers and seeds. The data creates a spatially registered record of what was applied to the field that the Cronins can compare with their yield data at harvest time. This helps them adjust their input use with variable seeding and fertilization rates for different parts of each field.

Looking out at the prairie, I felt like I was on a low-flying plane or a semi on the highway—only with a much better view. The unobstructed horizon-to-horizon vista was awash in browns and greens and the yellow-tan of native grass. For a moment I was almost convinced that I could see the curvature of Earth's surface. Birdsongs echoed across the fields and open sky. Cronin Farms is alive in every sense of the word—above- and belowground.

Mike, his eyes on the field, said that he likes to look two or three years ahead when thinking about how to improve soil health. For that he can blame Beck, who taught his brother Monty chemistry in high school. "It's because of him that people started no-tilling, mostly to get him to shut up and stop being such a pain in the ass."

ANOTHER ROUTE TO CONSERVATION

Curiously, the practices Beck advocates are not really all that new. Farmers around the world have long recognized the value of cover crops and crop rotations for enhancing soil fertility. Indeed, such practices were widely described in seventeenth- and eighteenth-century treatises on soil husbandry. But they were cast aside when cheap fossil fuels and chemical fertilizers ushered in agriculture's third revolution of mechanization and industrialization. What's new is combining them with not tilling.

Back when Thomas Jefferson sent Lewis and Clark up the Missouri River, he routinely grew cover crops and rotated crops—embracing two out of the three practices at the heart of modern conservation agriculture. But he plowed to control weeds. So did most of his contemporaries who practiced simple two-crop rotations, if they rotated crops at all.

Now decades after the Dust Bowl drove home the need to abandon the plow, we are going back to relearn the power of cover crops and diverse crop rotations. These practices worked long before the days of commercial fertilizers and pesticides, and South Dakota's success shows that they can prove effective on large modern farms in the developed world. What it takes is a new system based on combining no-till with the right mix of cover crops in complex rotations with the right timing—in other words, adopting all three principles of conservation agriculture.

As we drove back toward Pierre, I found myself wondering if this was the secret to reviving family farms across rural America. Was this the key to rebuilding prosperous farming communities, the original backbone of American democracy?

The fields along Highway 1804—as the Lewis and Clark trail is now known—testified to how much the landscape had changed since Beck began working in the region. Several decades ago, the region's soil regularly headed skyward after plowing. Now there was hardly any bare ground in sight as farmers went about spring planting. What did it take to so radically transform this landscape, which I had even featured on the cover of *Dirt* as the poster child for erosive farming? Adapting the old ideas of cover crops and crop rotations to the new idea of not tilling to forge a system that made economic sense for farmers and saved them work, through less plowing and lower input costs.

The primary reason for the initial decision to go no-till at Dakota Lakes was not erosion control, carbon sequestration, or any of the other advantages now touted for conservation agriculture. Rather it was the potential for greater moisture conservation and to save on labor and fuel. And when this led to higher yields using less fertilizer and pesticide, Beck embraced these added benefits.

He says that now the challenge is to keep moving forward. Policymakers and farmers alike have been assuming that progress means getting bigger—bigger farms and bigger equipment. Maybe that's not the right answer.

As he talked, I recalled his story of a local fourteen-year-old who had visited a tilled farm and was fascinated with how a plow turned the earth over—he'd never seen such a thing. To him no-till was normal. Many farmers in the region now consider tillage a catastrophic event, like a big earthquake or volcanic eruption. Imagine what could happen in the next thirty or forty years if no-till and cover crops become the standard by which *all* farmers run their farms.

It takes time to reestablish soil life and rebuild soil fertility. There is a transition period while converting from conventional practices, a few years of lower productivity and profitability. But Beck says that farmers struggle more if, when changing practices, they neglect one or more key aspects of the full system.

Beck's advice is simple: "Do your own cooking," he says. "Don't be afraid to ask for advice, but accept no recipes from others." He's found that there is no "best" crop rotation. It is a probability game. The specifics for any particular farm will differ from that at his farm or any other, due to unique soils, climate, and crops. But the basic laws of nature work the same everywhere, and he offers some ideas about the simple general principles that can improve farming systems:

1. Water use must match water availability, and no-till is more efficient in that it allows water to sink into the ground, where plants can get it.

2. Crop diversity and ground cover are necessary for fending off weeds, pests, and disease, so rotate cool- and warm-season grasses with cool- and warm-season broadleaf crops.

3. Crop rotations must not be consistent in either interval or sequence in order to provide the best protection against pests by preventing them from adapting; rotations should have a minimum of two years between the same crop.

I was struck by how Beck's ideas so closely parallel those Rattan Lal developed. Today, most subsistence farmers in Africa continue to use the traditional slash-and-burn practices. So I asked Beck whether low-input, no-till practices like his could work in Africa. He recommended I ask Kofi Boa, the director of the Center for No-Till Agriculture near Kumasi, Ghana.

7

DEVELOPING SOLUTIONS

When people are poverty stricken and starving they pass on their suffering to the land.

—Rattan Lal

As our plane came in off the Atlantic to land, the red soil created a striking contrast with the thick white clouds and the deep blue of sea and sky. From above, the city sprawled as far as the eye could see, a riotous collage of green vegetation, colorful metal roofs, and red-dirt streets. Landing in the early morning light, we rumbled past a jet without wings that had been abandoned at the end of the runway. I was surprised to find Accra, the equatorial capital of Ghana, ten degrees cooler than the unseasonably hot weather I'd left behind in Seattle. Even so, the humidity had me sweating before I got through Customs and past a seven-foot-tall sign blaring: WHAT YOU SHOULD KNOW ABOUT EBOLA.

Upon leaving the terminal, I instantly acquired a crew of new friends vying to offer me a taxi or point the way, for a small fee, to the domestic terminal—right next door. There, hanging out by a large electric fan, I enjoyed the spectacle of two middle-aged men performing an enthusiastic greeting—a booming hand-slap handshake followed by staccato bursts of popping finger snaps. Welcome to Ghana.

Staring out the window on the short flight north to Kumasi, once the heart of the Ashanti kingdom, I caught glimpses of red-orange stripes running across patches of grassland and forest beneath the thick, low clouds. From 14,000 feet, rivers and roads shared the same dirty color, distinguished only by their sinuous or linear form.

The flatness of this landscape reminded me of South Dakota, but I would find no rich black earth here. In this hot iron-red land, tropical rain had stripped the soil of readily accessible nutrients, leaving behind insoluble iron and aluminum. From its rusty color, I could tell that the soil didn't hold on to organic matter for long. Here nutrients accumulated in aboveground biomass, not below as in the prairie.

As my flight taxied in, I noticed that the façade of the terminal was the same color as the soil. A beaming twenty-year-old, Kofi Boa's youngest son Kwasi, greeted me and ushered me to a silver Toyota SUV. His compatriot from agricultural college, Kyei Baffour, fired up the engine and launched us out into Kumasi's busy streets as they peppered me with questions about sports, music, and Seattle.

We slipped into a languid stream of vehicles crawling through a sea of people wandering the streets with a variety of items for sale. Women toting boxes of fried plantains on their heads glided between moving cars, vying to catch the attention of travelers. I marveled at how, in all this chaotic activity, none of the precariously balanced goods ever fell. Each town we passed was a repeat of the same scene: a busy main road lined by small shops and open-air kitchens with roofs that matched the dirt.

Eventually we turned off the paved road, then drove several more blocks to my hotel. I settled into a dark second-floor room with no electricity, which meant a dead ceiling fan. Soon the mosquitoes drove me back outside to the crossroads, where the locals offered a steady stream of smiles from jam-packed, honking minivans.

As my lungs filled with that distinctive, peculiarly sweet burning smell endemic to the tropics, the roadside garbage—small corncobs and black, white, and blue plastic bags—almost looked festive. But when two passing trucks engulfed me in a black cloud of exhaust, I decided to head back toward the hotel. Before I got very far, a mesmerizing sight stopped me in my tracks: an iridescent, ten-inch-long African rainbow lizard, sunning its exquisite yellow, purple-blue, and gray-brown hide by the side of the road. No one else paid it any attention.

Back at the hotel, the Saturday night drums rumbled from the bar, pounding out rhythms that stressed every beat. This wasn't the 1-2, 1-2 of rock and roll. It packed a 1-1-1 kick, like Brazilian music.

The music went on late into the night. That and jetlag kept me awake, giving me a chance to reflect on why I was here. Over the past half century, most of the world passed through the "demographic transition" to the replacement level of 2.1 children per couple—zero net population growth. As of 2015, a hundred countries, including most of Europe and the Americas, had achieved this milestone. Now it's Africa that drives projections that we'll add another billion mouths to feed by 2050. Meanwhile, land degradation is growing too, and African agriculture faces substantial challenges for feeding both present and future generations.

Governmental and aid organizations generally tout the Green Revolution's fertilizer-intensive methods as the key to feeding Africa's growing population. But I'd done enough fieldwork in the rural tropics to know that subsistence farmers wouldn't be likely to participate in such a revolution. They don't have the money to buy high-tech seeds and chemical inputs. And big African farms geared toward world export markets aren't going to be feeding the hungry people at home. In regions around the world, introducing Western-style conventional farming with its high-inputs has pushed subsistence farmers onto marginal lands. What would a continued push in this direction mean for the smallholders who currently work 80 percent of Africa's farmland?

The generally acknowledged formula to end poverty is development through the investment of capital plus improvement of infrastructure and opportunities. This takes time. Development based on high-tech, capital-intensive farming won't raise rural living standards unless the infrastructure—like roads and new markets for farm products—is built out to match. It would also take decades—generations—to bring

African crop yields to U.S. levels, if indeed that is possible on tropical soils. So what can be done now to feed Africa's growing populace?

Most subsistence farmers in Africa still use variants of the traditional slash-and-burn method and a wooden hoe (known as an ard), developed more than 5,000 years ago. It's been shown that introducing modern mechanized tillage to tropical fields causes major soil erosion and degradation, yet adoption of no-till farming has stalled due to economic impediments like the high cost of herbicides. Farmers also remove crop residues and animal dung from the fields to use for cooking fuel. Reliably feeding Africa over the next several decades means that the continent's small-scale subsistence farmers need to become more productive, sustainable, and economically viable.

This was the challenge that brought me halfway around the world. Could conservation agriculture work on small African farms, the way it did on huge farms in South Dakota?

The next morning, Kwasi picked me up at the hotel and we headed off into traffic. I got lots of stares and enthusiastic waves, and even a rousing "Hey Mr. White guy!" as we crawled along past stores, stalls, and billboards advertising churches. At every speed bump, vendors stood ready to swoop in and sell water, food, or flip-flops. As we passed trucks burdened with enormous mahogany and sapele logs, I asked Kwasi where the forest was being cleared. He said that every day they bring more logs from the west—but not to worry, they won't run out. It reminded me of what some folks used to say in the Pacific Northwest, back when we still harvested old growth.

We stopped seeing logging trucks once we'd turned off the paved road toward Boa's village of Amanchia. As we began kicking up a trail of red dust, I admired the vines engulfing the telephone wires, and the lazy swooping of white-necked pied crows. The village proved a modest collection of earth, mud, and timber houses with open doorways and rusted roofs, built on red subsoil and weathered rock.

A little farther down the road we pulled onto a dirt track running behind a large No-Till Center sign. We'd arrived at Boa's farm, made up of small rectangular plots with various combinations of crops.

Stepping out of the car, I spied lots of quartz fragments on the ground. There's not a lot of nutrition in these deeply weathered rocks.

The first thing I noticed about Kofi Boa was his baseball hat emblazoned with: GOT DIRT? GET SOIL! The second thing was that he barely came up to my chin. He's about five feet three and sixty years old, with a dark complexion, striking smile, black square-top glasses, and a fringe of close-cropped gray-black stubble peaking out from beneath his hat. After welcoming me to the center with the casual ease of someone in charge, he wasted no time in starting our tour.

I followed him up the rough dirt track, past a series of small plots to an open-air shed, which Boa calls his classroom. All the fields we passed had one thing in common: no bare ground. The center occupies 4.5 acres, but Boa told me that he has another 20 acres of production plots that we'd be visiting the next day. It may be a modest operation by South Dakota standards, but it has already transformed the region around his village.

Boa set up the center to show subsistence farmers how to restore soil fertility. The first step, he said, was to get them to recognize the detrimental impacts of their traditional practices. And the best way to do that was to set up side-by-side demonstration plots. So, each year, he plants identical crops, using no-till methods alongside traditional slash-and-burn methods in neighboring concrete-lined, soil-filled plots 20 feet wide by 50 feet long. Each plot is set up to collect and display all the eroded soil in buckets and barrels at its downhill end.

The difference was striking. The no-till plot had a single 5-gallon bucket about three-quarters full of dirt. The slash-and-burn plot had three 50-gallon barrels full of dirt, more than twenty times the soil shed from the mulch-covered no-till plot.[18] Boa said this graphic display really gets farmers' attention. It certainly grabbed mine.

So did the difference in crop yields. Each plot was planted with corn followed by cowpeas (black-eyed peas). The corn harvest of the no-till plot came to 4.5 tons per hectare, whereas the slash-and-burn plot produced just 1.5 tons per hectare. The no-till cowpea harvest was 1.5 tons per hectare; the slash-and-burn plot produced 0.8 tons per hectare. In other words, the no-till plot produced triple the corn and twice the cowpea yields than the traditional plot.

Boa uses these plots to show how going right from slash-and-burn to conservation agriculture—skipping over the conventional Western combination of the plow and a lot of chemicals—makes sense here where organic matter breaks down so fast that the primary way to store nutrients is in the living biomass. His goal is to show villagers how to boost harvests with cover crops by doing it himself. Farmer to farmer—just like Beck's approach I thought to myself. Since everybody here knows him as a farmer, they figure if he can do it, they can too.

Boa established the No-till Center after farmer-philanthropist Howard Buffett visited in 2007. The story goes that when they met, Buffett was wearing a University of Nebraska shirt. Having done his master's degree there, Boa greeted Buffett with an enthusiastic, "Cornhusker!" They hit it off immediately, and Buffett was impressed with Boa's work, vision, and passion for conservation agriculture. Since then, Buffett's foundation has helped Boa to establish Africa's first no-till research center and demonstration farm by providing support for construction, equipment, and vehicles, as well as stipends and expenses for students. But like Dakota Lakes Farm, this is a working farm. Proceeds from harvests pay for day-to-day operations and activities.

It's a busy place and Boa is a very busy guy. In addition to farming, he offers several types of tours and classes for farmers—all for free. A one-day visit gives them a chance to see how Boa does things and to learn the concepts behind his approach. Those who get interested can come back for more. Farmers from the surrounding communities can take a course over consecutive Sunday afternoons, each covering a different topic: the concepts, principles, and practices of conservation agriculture; land preparation and planting; soil health and cover crops; and weed management. About once a month, Boa teaches the full course in a four-day intensive session for up to twenty farmers from afar. Students also visit the center during summers and holidays; more than five hundred came in the month before I did.

Finally, he also runs a fellowship program for six new graduates who have finished both college and their compulsory national service. They receive a stipend and free board for a year before Boa tries to place them in jobs. Boa's goal for these programs is to build a network of experts who can spread their knowledge and skill throughout the continent.

"When do you rest?" I asked him.

"I don't know rest," he replied. He's joking, but he means it. The group touring the center that day was the fourth in a week.

The practices that Boa teaches sounds a lot like Dwayne Beck's—no till, cover crops, and diverse rotations—adapted to a radically different topography, climate, and culture. How much of a difference has it made? It's more than doubled harvests and rapidly transformed his village. With farmers seeing good results within a year or two, most now own their homes instead of renting them. But best of all, reliable, larger harvests have brought food security.

MR. MULCH

I'd already noticed that the soil in Boa's fields was dark brown, a far cry from the orange-red dirt of the fields along the road. It might take years for a farmer to notice such changes in their own soil, but they notice reduced costs immediately.

A primary attraction of Boa's approach is that it reduces outlays for chemical inputs. He uses a fraction of the agronomist-recommended fertilizer and herbicide, and almost no pesticides. This is not to say that he has anything against agrochemicals. He considers them useful tools. But in his view the best way to promote their safe, efficient, and economic use is to minimize them.

He's found that once he's improved the soil, he doesn't get much of a boost from fertilizers anyway. Manure is cheap, but transporting and applying it is expensive. So Boa sees growing green manure on site— cover crops—as the best way for the local farmers to build soil. While alive, a cover crop that fixes nitrogen will help fertilize the soil, control weeds, and cut down on herbicide use. When dead, it rots and feeds the soil organisms that help build soil organic matter and soil fertility. In addition, some produce edible, marketable crops.

Another big selling point for Boa's methods is that with no plowing and less weeding, farming takes about a third less time. He is confident that once more farmers find that they can both save time and increase their yields, they will wonder why they ever used to plow at all. Plus, this efficiency gives them time to do other things. "I have good farmers who are also shopkeepers or full-time carpenters. How can they do this? Because of this system." This sounds to me like a step on the road to sustainable development.

Like Beck, Boa knows that he has to show rather than tell. "When farmers come here and see it, they believe it." And that, he says, is the point of the No-till Center.

When he started the center, the first farmers to come through complained that his test plots were too small. They wanted to see how his system worked on a full-scale farm—like Beck's farmers in South Dakota. And just as Beck is known among the Dakota Lakes locals, farmers here know Boa as one of them.

Boa grew up in Amanchia, the youngest of four boys. When he was twelve, a tragic experience set him on the path to becoming "Mr. Mulch," as he is widely known. One day, his widowed mother had not come home by dark, and nobody would tell him where she was. At 10 P.M. she finally returned, in terrible distress—her cacao farm, their only source of income, had burned. It was a total loss. Boa decided then and there to dedicate his life to fighting fire on farmland.

He knew that farmers used fire to prepare the land for planting, but given the recent fate of their farm, he started to investigate alternatives. That's when he learned about the *proka* system, a way of farming that the oldest men in the village used in their youth instead of burning the land before planting. In the local language *proka* means "allow whatever is on the soil to rot and come back."

When they wanted to farm a patch of forest land in the upcoming year, they would cut down the trees and chip up the brush. By the time they returned to plant, most of the vegetation would have decomposed. Through the residue they could then plant cacao or plantains. Boa absorbed the lesson. From then on he planted through mulch.

In 1991 he set off to the distant land of Lincoln, Nebraska, to study agronomy. "Why there of all places?" I asked him. Because he wasn't going to America for fun, he replied. At the University of Nebraska, no-till was a major research focus, and he was on a mission to get right in the thick of it. Now he counts himself one of the lucky few with experience in no-till on both large U.S. farms and small African ones.

After returning home, he worked at a research institute but quickly learned that he really enjoyed teaching farmers about how to change the way they farm—and change their lives. After the institute director said, "Hey Kofi Boa, remember: you are not an extension agent," he resigned. It was a risky, bold decision that he doesn't regret. To him, working with farmers is the whole point of applied research. "When I was thinking about leaving, I had people tell me I was crazy. But I knew that if I never got anybody to work for, I could work on my farm." A few years after he began using his farm to teach other farmers, Buffett helped turn Boa's dreams into the No-Till Center.

It was almost 2 P.M., and I was about to see Boa's version of Sunday school in action. But first he wanted to show me a couple of things.

He led me from his open-air classroom back to the end of the dirt track off the main road. In the field beside the parking area, he was conducting an ongoing experiment in growing corn with and without organic fertilizer. Notice, he said, that there isn't much difference in the height of the corn, even though the plants on the right side received 3 tons per hectare of organic manure, whereas those on the left were grown without any. The fertilized crop was slightly darker green, but the unfertilized crop was still very healthy. Both towered over Boa as he waded into the field. "If the soil is healthy, you don't need much fertilizer," he said.

"So how do you keep it healthy?" I asked. You can probably guess his reply. "Mulch!"

Next he showed me a contraption that he used to teach farmers about the effect of cover crops on soil moisture. It's an important issue here, because streams are few and far between and there is hardly any irrigation. Rain is all farmers have to work with. They need to get it down into the ground—and keep it there until crops can take it up.

Boa led me over to what he calls the Amanchia mobile soil-loss demonstration kit. There, sitting out by a field, was a tabletop version of his full-scale erosion plots. On a green metal rack were three side-by-side, several-inch-deep trays of soil. The one on the left consisted of bare plowed soil, the middle one was no-till soil covered with a layer of mulch, and the soil on the right was traditional slash-and-burn, with the organic matter torched from the surface. At one end of each tray, the metal walls angled inward to form a constricted outlet, under which jars sat on a flat ledge to catch the runoff that spills through after rainfall lands on the trays.

The difference was striking. The jars below the plowed and slash-and-burn soils held what looked like chocolate milk, opaque and thick with dirt. Weeds sprouted in the bare plowed and slash-and-burn trays, both of which were dry, crusted, and cracked at the surface. The water in the jar below the no-till soil was yellowish with dissolved organic matter, but clear enough to see just a few grams of soil lying at the bottom. No weeds were coming up through the mulch in the middle tray. The soil below was moist.

Boa didn't have to tell me anything more. He just stuck moisture and temperature probes into the trays. The bare trays had 9 and 12 percent moisture and surface temperatures of 99°F and 108°F (37° to 42°C). The no-till tray in the middle had 30 percent soil moisture and a temperature of 87°F (31°C). It was a remarkably simple—and effective—demonstration of how leaving mulch from a cover crop can keep the soil more than ten degrees cooler and retain three times as much water where crops can use it.

When Boa headed back to the shed to prepare for class, I wandered the farm to look at his soil. His fields had close to complete ground cover blanketing dark brown silt and clay. The soil held together well, yet I could crumble it with my fingers. It was a far cry from the hard-baked subsoil exposed along the road.

Soon seventeen men and two women arrived and converged on the shed to take their seats in blue and red plastic chairs beneath the corrugated metal roof. They ranged in age from twenties to sixties; some in sandals, others in shiny black leather shoes. Some wore dirt-covered shirts, others their Sunday best. Two were wearing colorful African prints and two were in elegant long-sleeved gowns.

Boa stood at the front of the classroom in a gray T-shirt, brown slacks, tennis shoes, and his GOT DIRT? GET SOIL! hat. He started with a prayer. Then he launched into the subject for the day: weed control. Kwasi translated from the local language for me.

An animated speaker, Boa gestures with both hands in big sweeping arcs and asks questions like Socrates—why do we till, why do we spray herbicides? It took less than a minute for me to see that he's a naturally gifted teacher.

Behind him on the table was a row of twenty herbicides in brightly colored plastic containers. Boa explained what was in each—glyphosate, atrazine, and 2,4-D. "Some are systemic and kill the whole plant, but others just kill the green parts," he explained. Remember, he intoned, a chemical that kills to the seed level can impact cash crops if it lasts in the soil. And farmers often use too much herbicide.

"I have seen a lot of people arguing that no-tillage is very dependent on herbicides—I hear that everywhere—but in an environment like ours it is not true." A typical farmer will slash and burn and leave the ground naked. And weeds are sure to come with the first rains because on bare ground they have no competition.

"So they invite the weeds to come and then go and spray with an herbicide like glyphosate. But when they do that, they then need to follow up with another herbicide, paraquat or atrazine." In contrast, leaving mulch on no-till fields has helped farmers control weeds enough to at least halve herbicide use. Boa has even found that there are places in the spring where no herbicide is needed because crop residue keeps the weeds down enough for an emerging new crop to elbow them out.

Cover crops are another way to fend off weeds. Properly used, they allow farmers to reduce the amount of herbicide and eventually eliminate it altogether. With leguminous vines as cover crops, a farmer can use a cutlass (the local name for what I know as a machete) to chop the plants at ground level and leave them to rot as mulch, providing a smallholder with cheap, quick, and effective weed control and fertilization.

Heads nodded when Boa added, "Those making the chemicals want the farmers' money. Using cover crops to control weeds means you don't need to use so many chemicals. You keep more of your money." The point of all this, he went on, is not to stop using herbicides and fertilizers altogether, but to abandon the "more on" approach—the idea that if a little is good, more is better. The goal should be to use as little as possible to produce an abundant crop.

It was time to leave the classroom for a tour of the fields. The students and I followed Boa as he relayed the history of each plot. None are planted in the same crop twice in a row and, given the year-round growing season, they can get several crops annually if Boa and his crew plan the rotation correctly. Boa makes the call based on the condition of the ground after each crop and on market conditions—what he will be able to sell at the time of year the crop will mature.

But there are other considerations too. A deep-rooted crop should follow a shallow one. High-biomass-producing crops should follow low-biomass ones. And a nutrient fixer should follow a nutrient scavenger. In other words, there is a pattern and rhythm to crop sequencing.

Explaining all this, Boa used his cutlass as a pointer, like a natural extension of his arm. Then, perhaps for my benefit, he showed us how to plant and fertilize with a cutlass. It involved a quick jab of the cutlass, a flick of the wrist to pry soil up, and a dropping of a seed into the resulting hole. This was followed by a quick tap of the cutlass to pat it all back down, break up the surface crust to help the seed come up, and establish good soil to seed contact. Then, to sparingly apply fertilizer, he pushed the tip of his cutlass into the soil, this time beside a one-foot-tall corn plant. He pried up the soil, dropped a bit of fertilizer from his other hand into the hole, and used the cutlass to pat it back into place. This is low-tech, low-capital precision agriculture.

Throughout this demonstration, one farmer squatted, compulsively pulling what few weeds he could find from between corn plants. Taking this in, I marveled at how anyone could think that a high-tech, fertilizer-intensive approach with costly GMO seeds would be practical for these farmers. In reality, they need advice on how low-input, low-cost practices can best increase their harvests.

Boa summed up his philosophy: "What we do is not organic agriculture; what we do is work with nature. If we have a serious problem with an onion field that needs a fungicide, we will use it. But we don't need to use much herbicide or pesticide."

As far as I could tell, nature likes to work with him too, judging by all the butterflies and birds on his farm. Some cover crops flower at times when other plants don't, so a diverse rotation helps maintain pollinators. He said, "The birds have made this place their home." Their songs filled the air as Kwasi and I drove off, and I began to wonder whether organic-ish farming might change the world—and save agriculture from itself.

THE ONLY BARE GROUND IS THE TRAIL

That night, back at the hotel, the power was still switching off at random, rendering the fan useless. I slept in fits and starts in the sweltering heat. When I wasn't dozing, I was worrying about the lack of mosquito netting in this malaria-prone country. The next morning, I broke out in a sweat over breakfast.

Kwasi and I stopped off at the No-Till Farmers House in Amanchia on our way to the Center that morning. It's a residential compound where visiting farmers stay during training sessions. As we pulled in, a woman walked by with a basket full of foot-tall seedlings on her head. When I asked Boa about it, he explained that she carries the transplants to her field one basket at a time. On Saturdays she enlists her kids to carry more since they're not in school. How, I wondered, could Green Revolution technology help her? Free drought-resistant crop varieties might; expensive or fertilizer-hungry ones won't.

We headed off to the Center's production fields to meet two busloads of students from a regional agricultural college. We parked where Amanchia gave way to fields, next to a pedestal of soil and rock several feet high, where the dirt road had been carved through town. The roadcut exposed a half-foot of true soil above a quartz vein running through weathered rock. This thin layer of topsoil would wash away if exposed to rain, and its fertility was based on rapid turnover of organic matter. But I was about to see a full-scale demonstration of how Boa kept topsoil both in place and fed with organic matter grown on-site.

Once the students had arrived, Boa led us on a long procession down a path between fields and out into the countryside. The whole landscape was divided into planted fields, with no space between adjacent farms and their great varieties of crops and ground cover—or lack thereof.

Our first stop was a citrus plantation across the track from Boa's production plots. He instructed the students to look at the soil beneath the trees. It was bare, hard, and crusty, the topsoil gone and the red-orange subsoil exposed at the surface. There was no organic matter and no soil structure, but plenty of weeds. He said that the lady who owns this field never harvests much.

We crossed the road and Boa walked us into a field of corn and plantains, with an understory of cowpeas and bushy mucuna. "You want to use cover crops," he said, pointing out the soil-enriching, low-lying legumes living harmoniously with the taller corn. "Planting cover crops here we didn't need to do any weeding."

It's important to use several cover crops with different attributes, he explained. Cowpeas are a shade-tolerant legume, an ideal staple for interplanting with other crops. Jack beans, lablab, and mucuna are also legumes, but they grow as vines that can climb cacao and plantains. "This makes them easy to terminate. You just chop them with a cutlass. It goes real fast." And once they're dead they rot and enrich the soil.

Last season, Boa tried an experiment on two side-by-side plots after planting plantains. He didn't need to weed the plot where he had also planted a cover crop of jack beans. But right next door, he didn't plant anything between the plantains, and so far they had to weed that plot three times and spray glyphosate. The lesson? Let the biology work for you and save time and money.

"Above all," Boa continued, "keep the soil covered." You want to create a situation where you can dig into the ground with your bare hands. When you have cover on the ground, it fosters the biology that tills the soil for you, and an organic matter cover reduces runoff so that water can sink into the soil. He reminded the group of the capillary action of the soil, which draws water up from the subsoil into the upper soil. What happens when we plow? It breaks the capillary pores and subsoil water can't rise to the surface—and the plants can't drink. And because organic matter doesn't conduct heat very well, less evaporation happens in the soil if you keep the ground covered.

Boa says that while this year is the driest he can recall, all his pepper seedlings survived. Yet farmers who plowed lost all of theirs. Why? When they plowed and broke up the soil, they lost soil water. If you don't work the soil, it will stay moist longer and help plants survive.

To support a healthy crop even during times of drought, he advised the students to pay attention to three things:

1. Open up the soil only enough to plant a seed—minimally disturb the soil.

2. Make sure the ground is covered with prior crop residue at planting, and use cover crops to build biomass and soil organic matter.

3. Introduce diversity to the fields through rotations and intercropping.

I'm sure you recognize this list. Here, again, are the principles of conservation agriculture.

We continued on into the surrounding fields, Boa guiding a train of eighty students through the production plots along a single-track trail. It felt like hiking through a food jungle. Alternating rows of broad-leaved plantain and cacao were interplanted with corn and an undergrowth of cover crops including colorful orange peppers. Cacao is the best cash crop, but it takes years for the plants to mature and the young plants need shade, which the plantain provides. In the meantime, the corn, peppers, onions, and cowpeas help build soil and provide another source of income. This polyculture minimizes inputs and increases drought resistance. The only bare ground is on the trail.

The last stop at the edge of the production plots was at a field recently cleared of forest to prepare for planting. Boa stopped and gathered the students around him into a half-circle. "See how all the slash is left on the field and just the big wood was removed?" Leaves, branches, and chopped-up brush covered the field. "If you talk to very old people," he said, "you find that this is what they did before mechanization. We are going back to their system, but with a scientific mind—doing what the old people were doing but with new knowledge."

Boa contrasted this scene with the slash-and-burn approach (farmers burning off all the organic matter) and the Green Revolution approach (agronomists clearing the fields down to bare earth before applying fertilizers). "Without you ever coming here, would you have believed it?" A chorus of "no" rose from the students. I was glad to see I was not alone taking notes in this group; almost all of them were too.

We returned to the shed, and Boa continued his lecture, showing us photos of forest and of bare land. "We can all see what deforestation does, but one thing that people overlook is the destruction of the soil by our own farming practices—what everybody is doing. Think about the forest, nobody does any digging there, but it is very productive. Look at these photos! Which land would you want to farm if you had a choice?" A student spoke up, saying he'd choose the forest. Boa asked if the class agreed. Everyone did.

"Why?" he asked.

"The forest soil is moist," one student offered.

"The forest soil has more organisms—more microbes and worms," said another.

"Yes," Boa added, "to cultivate soil organisms, we must create a shaded, moist environment. Soil life can't take the heat of the African sun. And they need the litter on the soil surface for food. See what has happened? All of us would choose the forest because if you put it all together we see it as more productive. A farmer's goal," he concluded, "should be to replicate on their farms the conditions that make the forest productive."

Boa explained another benefit of conservation agriculture, highlighting a point especially relevant to smallholder farmers like those watching him. When a farmer grows a diversity of crops using rotations, associations, and intercropping, it's like a do-it-yourself insurance policy. If one crop is lost, there's another one to fall back on. If 90 percent of a farmer's land is planted with a cash crop, like ginger, then the remaining 10 percent should be planted with food crops in case the main crop fails.

Boa explained how the trick with cover crops is to get a mix of nutrient scavenging crops to gather and incorporate nutrient elements from the soil into their biomass. Deep-rooted cover crops bring nutrients up to the surface, and even nonlegumes return nutrients to the soil when they die. So, if you have too much compost and no place to store it, you can put it on the soil and plant a cover crop to capture and temporarily hold these valuable nutrients. Then when you want to use the nutrients, you cut the cover crop and let it rot, returning nutrients to the soil.

In the tropics, soil organic matter is naturally very low. The hot, wet environment enhances microbial decomposition of organic matter. Boa says that while soils in Ghana's forest reserves have up to 4 percent organic matter, most croplands now have only 1 percent, at most, due to decades of plowing and continuous slash-and-burn. Though the ash does provide nutrients, often it gets washed away by runoff or blows away in the wind. Boa sees slash-and-burn as feeding the soil organisms for a day and starving them for a week. The key to building soil organic matter over the long run? Produce and retain biomass, exactly what cover crops and crop residue accomplish. In Boa's experience, integrating cover crops grown for nutrient capture and release as part of crop rotations can increase soil organic matter levels from around 1 percent to almost 2 percent in less than two years. He thinks that over decades his approach can push his soil up above 4 percent organic matter, higher than what is found in the forest.

When a student asked whether he uses fertilizer, Boa responded with a story. If you have a barrel at home and it overflows after a downpour, would you ask your brother to go fetch water to top off the barrel? No, you wouldn't, because the barrel is full. But if the barrel is only half-full after the rain, you would send your brother to fetch water to top off the barrel. The soil is like that barrel. With the return of organic matter through decay of crop residue, you can keep it full. Of course, you can also top off the fertility with fertilizers—but you won't need much if you have a full load of organic matter.

Over the course of many experiments, Boa has found that farmers who convert to no-till reduce their fertilizer use by at least half, more if they integrate cover crops. In some of his plots, he finds he can reduce fertilizer use more than 90 percent.

Boa added that if he does use fertilizer, he uses organic fertilizer. But he suspects that within a decade, Ghanaians could go completely off chemical inputs if they integrated all his practices to build soil organic matter. This would help restore the mycorrhizal fungi and bacteria that scavenge phosphorus from the soil. Throughout much of Africa, the total phosphorus concentration in the soil is adequate, but it is almost all locked up in insoluble minerals. Soil life can turn inorganic phosphorus into forms that can enter biological circulation. Unlike soluble mineral fertilizers that provide a quick splash of nutrients, accompanied by high rates of leaching, mycorrhizal fungi provide lower amounts stretched out over time. Recruiting and keeping mycorrizal fungi in one's fields is another reason not to plow. As Boa puts it, "Mycorrhizae are so small and fragile," he said. "Why would we disturb them?" Instead, a farmer should feed them. And what do they eat? Organic matter.

Mineral fertilizers provide a lot of major elements, but most lack micronutrients that plants also need. Organic matter has both the micronutrients and the major nutrients. Boa relates the importance of micronutrients through a cooking analogy: "Think about when you are making a soup. You use a lot of water and a lot of meat, but just a little salt and pepper. Few people like to eat soup without salt and pepper. You don't need much, but it's important to have it in there."

PROOF IN PRODUCTIVITY

Beneath the hot African sun, water can be scarce and droughts are a growing problem. Boa showed me an ingenious drip irrigation system made from discarded 1.5-liter plastic water bottles, of which there is no shortage. He pokes a small hole in the side of a bottle's bottom, fills the bottle with water, and then places the perforated, leaking vessel into a small depression at the base of a cacao tree. This delivers excellent drip irrigation right to where the plant needs it. Farmers love this free "technology," because it uses something they consider waste.

Boa told me more about his experience over a late lunch of onions and a cooked mash made from elephant-ear sized cocoyam leaves, scooped from a wooden bowl using our fingers and plantains as utensils. "Everything came from the farm," he said, pointing to the plantains growing twenty feet away. A decade ago, people called him crazy; now almost all the farmers in town use his system.

For dessert, there was cooked corn on the cob, also from the farm. The smallish cobs of blackened kernels were nutty and filling. Looking around, I asked Boa where he wanted me to put the bare cob. He replied, "Anywhere, just don't waste it." I promptly tossed it back into a cornfield, and he laughed approvingly. A few years back, Boa began telling farmers in his village, "Whatever doesn't go into the tummy goes back into the fields." Now most of the time when he sees someone carrying a basket out to the fields, it has household organic waste in it. Whatever they take to the house but don't eat, like plantain or cassava peels, they return to their soil.

After the students packed it in for the day, Boa, Kwasi, and I headed out to visit a couple more farms. At the No-Till Farmers House, we met up with Adu Menzah, a farmer whose gray Old Navy tank top matched the color of his stubbly beard.

To get to his farm, we walked through the village of mud-walled houses, most with laundry hanging outside to dry. On our way past one house, a curious matron charged up and, after a thorough inspection of me, declared, "I love you." Boa chuckled as I assured her the feeling was mutual and we continued on to the fields.

As we followed the single-track trail, Boa offered running commentary, starting with expressions of dismay at the bare-ground fields of tilled tomatoes and corn, planted by recent immigrants who'd sprayed herbicide beforehand. Notice, he said, how their tomatoes now have fungus. Next, he pointed to a field greener and taller than the neighboring one, with plenty of decaying organic mattter on the ground. At the next field, he showed us the prevalence of worm castings where crop residue covered the ground. He poked his cutlass into a no-till field and revealed the soil to be moist and full of worm burrows. When he tried the same thing in the neighboring traditional field, the soil was dry and hard-packed, with no sign of worms.

After walking about a mile through fields with various combinations of cacao, plantains, cassava, corn, and cover crops, we arrived at Adu's farm. He told us that he had started with no-till and cover crops a couple years earlier. Although he did not use any fertilizer or herbicide, he had already seen a lot of change. Despite a very dry year, the plants on his no-till fields were doing better than his other fields. So he'd decided to follow Boa's example and convert his whole farm. He'd interplanted cocoyam along rows of plantain, then planted rows of taro and cassava in between the plantain rows. He also had yams growing up on poles and some snags he'd left standing from when he'd cleared the forest. Beneath it all, an underplanting of cacao was starting to poke up. The other crops created a good microclimate for the young cacao, which would eventually become his main cash crop. They also provide food and income until the cacao matures. Adu had six crops growing simultaneously in this field. "This is what I want to see!" Boa declared.

Adu had decided to stick with no-till and convert his other fields based on one key incentive: the difference in yield. Of course, he also liked saving on fertilizers and herbicides, as well as the fact that his fields are now full of birds, especially his favorite, the small songbirds.

He estimated that his net yield per unit area had quadrupled, which is at the high end of Boa's estimate of a three- to fourfold difference for years with bad rainfall. When I remarked that four times the yield is better than the yield increase from the Green Revolution, Boa laughed heartily. "Yes, much better!"

Together we returned to the No-Till Center, and Boa introduced me to Kokor Yaa, a young farmer dressed in jeans, rubber boots, and a long-sleeved flannel shirt—an outfit I couldn't imagine wearing in that heat. But she looked perfectly comfortable as she led us off on a trail into the fields. Crossing over a series of rolling hills and valleys, I saw no surface water, no streams. The trail reminded me of walking through jungle in the Philippines or Amazon—but here everything was planted.

After bushwacking through cover-cropped cornfields and cacao forest for what felt like a mile, we came to a well-defined footpath that took us through a several-year-old plantain field with bare ground. Boa said to no one in particular, "Who did this? Oh, boy, it looks naked."

We arrived at Kokor's field, a rectangle with peppers at one end and plantains at the other. She's never used any pesticide, for the simple reason that she can't afford any. Because she'd slashed and burned this field for more than a decade before going no-till, the ground was covered with plant residue but still contained a lot of charcoal in the soil. Why did she change her methods? Boa's Sunday school opened her eyes. She likes no-till because "even if it doesn't rain, the soil is moist." That, and the time she saves. It's a long walk to and from the small shop in town where she works after she tends her field.

On the way back, Boa moved through this food jungle with steady sureness. He looked at every patch of crop, asking aloud: Is the soil covered or naked? Is it moist or dry? Are the crops tall and green or short and yellow? As we stepped onto the road back in front of the No-Till Center, Boa told me that he's pleased with what he saw. "This year has been very dry, a good test of no-till."

By this time, I was drenched in sweat. Thankfully, Boa's chief farmer brought out a bag of green coconuts and a cutlass. He swiftly trimmed off the outer coconut shell, then cut a window into the top and pried it open so that we could get to the refreshing water inside. I'd been sweating so much that Boa insisted that I have a second. I was in no condition to object.

YOU CAN'T EAT GOLD

Once I'd finally cooled down, Boa took me to see one more thing: the destruction of the fertile valley bottoms by mining activity in the neighboring district of Amansie West. As we drove away from Boa's village, the country became a hilly land of red and green—red soil and houses, green forest and crops. But the farther we went from the No-Till Center, the more bare, plowed ground we passed, each naked plot lacking in discernable organic matter and topsoil, with just reddish subsoil exposed at the surface. In some of the villages, it seemed like every house had a little store, and tangles of sheep obstructed the streets. Soon it became clear that although sheep have the run of towns, mining companies have the run of the land.

Over the past decade, Boa told me, Chinese mining companies have moved in to dig for gold in the river gravels that form the valley bottoms. They sieve the gravel, take the gold, and cast the leftovers back into long windrowed piles of tailings. Entire valley bottoms have been torn up and turned over.

The physical disturbance has not only destroyed the soil, but the spoil piles leach toxic runoff. As we crossed a putrid, muddy stream, Boa recalled, "When I was young, the streams here were crystal-clear; now you can't drink from them." In the mining district, villagers who used to get their water from streams now have to buy it in plastic bags from trucks that deliver daily to their village.

What was once good farmland is now a spoil-pile moonscape. Originally, no one could own the valley bottoms; they were protected by village chiefs. But the mining companies offered a lot of money, and one chief after another decided to take the quick cash and allow the destruction of the land and water. According to Boa, this style of shortsighted mining is starting to happen all over Ghana. He worries that governments across Africa are so hungry for investors, they don't always care what they do. "The valleys are the most fertile lands. This was good land. This is so sad. The death of the land."

The next morning before I had to leave to catch my flight, Boa and I sat around an orange plastic table with matching chairs on the upstairs veranda of the No-Till Farmers House. He smiled and spoke slowly as I wrote quickly, struggling to keep up in my attempt to take down his parting words. From our perch in Amanchia's only two-story building, we could see a panorama of the village, its mud-brick houses and its inhabitants walking by, balancing loads on their heads, from flip-flops to fruit and chainsaws.

As he spoke, Boa's eyes shone with evangelical passion. He is like a prophet of no-till in Ghana, trying to lead his people through several agricultural revolutions in a single generation, to go straight from slash-and-burn to conservation agriculture.

I asked if conservation agriculture can feed Africa, and Boa quickly replied. "The blunt answer is yes; we will especially need it as climate change continues to stress food production. Africa has traditionally been fed by small farms. If we combine no-till with appropriate crops and systems, I don't see why we can't continue to do that."

People have been coming to this center from all over Africa to learn how. The productivity of subsistence farms in the continent is low, and it's been declining in recent decades as fallow land isn't available and soil degradation is widespread from continuous cropping with traditional methods. There is substantial interest in whether changing high-disturbance slash-and-burn practices to low-disturbance methods using cover crops, intercropping, and crop rotation can push progress toward sustainable development.

Boa sees low input costs as central to making conservation agriculture an effective way to address rural hunger and poverty among smallholders in Africa. He'd shown me how the same principles that transformed South Dakota can increase, and in some cases more than double, harvests for subsistence farmers—all without expensive irrigation or planting equipment and with minimal expense for inputs like fertilizer and herbicides.

Yet there is controversy over the potential for conservation agriculture to feed Africa, as reviews of no-till and conservation agriculture have come to conflicting conclusions. Some studies have reported that conservation agriculture required substantially more inputs or produced lower yields. I suspect that part of this is again due to variability in how conservation agriculture practices are defined and implemented. A 2014 study that reviewed available literature came to the conclusion that few studies had evaluated the effectiveness and cost of conservation agriculture when *all three* of its central principles were adopted.

One that did, a meta-analysis of sixty-one studies of sites across sub-Saharan Africa, found that increased crop yields under conservation agriculture depended on, again, adoption of all three aspects of conservation agriculture—no tillage, cover crops, and complex crop rotation. Such comparisons are complicated, however, by the reality that even if the principles are universal, the specific practices that work best will vary regionally and with farm-specific contexts.

Like Rattan Lal, Kofi Boa bemoans the wide range of practices called no-till. Why, he asks, would anyone even talk about, let alone do or study, bare-ground no-till? It just makes no sense when "bare soil is worse than tillage."

There is no question that soil erosion must first be curtailed before soil fertility can be built up to sustain higher yields with fewer inputs. In the late 1970s, experiments in northern Ghana found about a twenty-fold difference in erosion rates between bare and mulched fields. A more recent study of conservation agriculture in Zimbabwe found that coupling the use of organic matter as soil cover with integration of legumes into crop rotations not only reduced erosion, but helped reverse declining soil fertility and increased soil organic matter. Average corn yields under conservation agriculture were consistently higher, up to double those of nonconservation agriculture fields. And a long-term field trial in Zambia reported a 9 percent increase in soil carbon content in conservation agriculture fields, while there was a 7 percent decrease in neighboring conventionally tilled fields. In rural Mali, the combination of conservation agriculture with limited fertilizer use, or microdosing, provided short-term yield benefits combined with longer-term building of soil quality and soil organic matter. The take-away from these studies seems clear: here conservation agriculture builds soil, and conventional agriculture degrades it.

A number of studies in sub-Saharan Africa have reported better economic returns for conservation agriculture than for traditional practices. Even when similar or higher yields are obtained, there can be significant savings in labor and time. For example, a three-year study of twenty-four farms in Malawi found that conservation agriculture plots with intercropped maize and pigeon peas (*Cajanus cajan*) produced a third higher maize yields after the first year, in which there was no difference. It also took farmers just three-quarters of the time to manage, and was twice as profitable as conventionally tilled maize. The soils on these farms had less than 1 percent organic matter, a value considered a minimum for maintaining crop production in the region. The study concluded that agronomic practices that increase soil organic matter were critically needed.

In another review of studies of reduced-tillage and no-till practices across Africa, the authors found that mulching was the key to improving soil properties and increasing crop yields. And a curious experiment in northern Burkina Faso found that mulching previously degraded soil attracted termites that devoured the mulch, breaking it down and thereby recycling nutrients and helping restore fertility to the land. The researchers applied several tons per hectare of dry straw and shrub mulch to completely bare and crusted soil, which had been referred to by local farmers as *zipella*, or "dead soils." Soon termite burrows perforated the surface crust and allowed water to once again sink into the ground instead of running off.

Only a small number of smallholder farmers in sub-Saharan Africa have adopted full-on low-input conservation agriculture like what Kofi Boa teaches. A key limitation to expanding the adoption of cover-cropping systems in Africa is the availability of cover-crop seeds suited for African conditions. Boa said that farmers constantly ask him for cover crop seeds to help restore land they say has "spoiled" and is no longer fertile. They've learned what cover crops can do, but where can they get seeds? It looks like they'll soon have a source. Howard Buffett's foundation has converted parts of his farm in South Africa to produce high-quality, low-cost cover-crop seeds for distribution across Africa.

When I asked about other obstacles to broader adoption of conservation agriculture in Africa, Boa rattled off a short but daunting list: tradition, lack of political support, bad policies, and the difficulty of gathering and disseminating knowledge. African governments and foreign aid programs tend to focus on input-intensive approaches behind the get-big-or-get-out philosophy that transformed American agriculture in the late twentieth century. While the approach Boa advocates makes short-term economic sense for subsistence farmers, implementing it involves a change in thinking, rather than using products made by the companies that dominate the global business of agriculture.

This thought lingered with me as we drove to the airport, past small roadside stalls selling old-school treadle Singer sewing machines and racks of chickens in cages, a woman carrying a stack of ten egg boxes—sixty dozen eggs—balanced on her head, and a wrecking yard assembling rainbow Frankenstein taxis, each panel a different color. As I readied myself to leave Africa, I couldn't help but think that the way for aid organizations and philanthropists to have lasting impact there is not through encouraging the "more on" approach, but through practices like Boa preaches. And, as he's shown, the way to do this is to use demonstration farms to bridge the divide between research and local experience, between science and practice, to build a foundation of knowledge relevant to smallholders.

Despite the huge differences in soils, climate, and farming practices, the same guiding principles worked in both South Dakota and Ghana. Dwayne Beck patterns his practices on nutrient cycling through semi-arid grasslands. Kofi Boa patterns his on nutrient cycling through tropical forest. Though specific practices differ depending on the situation of the farmer, I was starting to think that soil-building farming could be done around the world.

Both Dwayne Beck and Kofi Boa use herbicides and fertilizers, albeit sparingly. But can no-till work in conjuction with organic farming at scale? Can expanded cover crop and legume cultivation also improve organic methods? To explore these questions, I decided to visit the longest-running test of organic no-till farming, near Kutztown, Pennsylvania.

8

THE ORGANIC DILEMMA

If herbicides are so good, how come we still have weeds?

—Gabe Brown

People tend to assume that organic farming and sustainability go hand in hand. But that's not necessarily the case—and it hasn't been for most of history. While going organic has some big advantages, even today most organic farmers still rely on the plow—the chief culprit in this story. Why? Because it provides cheap, reliable weed suppression. Yet, as we know, it's not the only option, nor are herbicides always a better alternative.

As far back as 1938, USDA agronomist Clyde Leighty wrote that "Rotation of crops . . . is the most effective means yet devised for keeping land free of weeds."[19] He suggested that the best defense against weeds was a cover of beneficial plants, grown to outcompete weeds and provide green manure.

Leighty also knew that crop rotation boosted harvests. "There are numerous cases on record to show the advantageous effects of rotating crops on yields and quality . . . the ravages of insect pests and of plant diseases are reduced to a minimum in a crop-rotation system . . ."[20] To back up his claim, he presented data from a thirty-year field trial at the Missouri Agricultural Experiment Station, which showed that a four-year rotation of corn, oats, wheat, and clover increased yields by two-thirds to more than double those from monocultures.

Farmers had known about using cover crops to enhance soil fertility and bolster crop yields for centuries. But these practices lost their luster after the Second World War, as plentiful supplies of cheap chemical fertilizers offered comparable boosts in yield and allowed farmers to specialize in particular commodity crops and eliminate all the messy bother of animal husbandry. In short order, agriculture came to rely on chemical fertilizers to support high-yielding monocultures. Using cover cropping and crop rotations to maintain or build fertility and suppress weeds didn't sound like progress when fertilizers and herbicides proved cheap, easy, and effective—at least in the short run.

That, however, wasn't the only factor at play. Chemical manufacturers had influential allies in governments interested in supporting fertilizer factories that could convert back to munitions production on short notice.[21] While not everyone jumped on the agrochemical bandwagon, increasingly specialized academic, industry, and agency researchers tended to dismiss organic farming as a throwback to pre-scientific days. By the 1950s, the USDA had gone all out to champion agrochemical farming.

No-till farming became more attractive after the advent of herbicides that offered a simple, practical alternative for weed control. Interest ramped up even more with the introduction of crops genetically engineered to tolerate the herbicide glyphosate, which kills nongenetically modified plants by disrupting a basic physiological process. The effectiveness of glyphosate gave its manufacturer, Monsanto, a huge advantage in the herbicide market and a monopoly on patented seeds for glyphosate-resistant crops. Conventional farmers loved glyphosate, even if using it didn't appreciably improve per-acre harvests. They hated weeds and it killed them really well, offering easy, efficient weed control without plowing—at least at first. This, in turn, helped drive widespread adoption of genetically modified corn and soy.

The first herbicide-resistant weed surfaced in 1970, an atrazine-resistant groundsel. Six years later, Monsanto registered glyphosate for broad-spectrum weed control. The first glyphosate-resistant weeds, Wimmera ryegrass (*Lolium rigidum*), were reported in Australia in 1996—the year that glyphosate-tolerant crops were first introduced. By 2014, almost two dozen weed species with glyphosate-resistance were reported in the United States. Worldwide, a total of 432 different types of weeds were resistant to various herbicides. Naturally, this renewed interest in whether all those herbicides were really necessary after all. Could no-till be done organically, or did one have to choose between herbicides and the plow?

There's no better place to ask this question than at the Rodale Institute, a 333-acre farm near Kutztown, Pennsylvania. The institute's namesake, Jerome Rodale, was one of the first American advocates for organic agriculture. The sickly son of a Jewish New York City grocer, Rodale had built a successful manufacturing business, but in 1930 he decided to move his family to a farm in Emmaus, Pennsylvania.

When he asked university scientists what he should do with his newly purchased 70 acres, they recommended using chemical fertilizers and pesticides. This didn't strike Rodale as particularly good advice, given the fact he'd left the city in the first place so his kids could have a healthier environment to grow up in. After reading Sir Albert Howard's ideas about the relationship between soil health and human health, Rodale linked the word "organic" with agriculture and started an experimental organic farm in 1940. Two years later, he started *Organic Gardening and Farming* magazine (now called *Rodale's Organic Life*). Publishing books and magazines about health-related issues made Rodale famous, and he became a prominent advocate for healthy living and organic farming using cover crops, crop rotation, and manure. Until, that is, he died of a heart attack on the set of *The Dick Cavett Show* in 1971. That year, Rodale's son Robert purchased the Kutztown farm.

I'd seen Jeff Moyer, the institute's farm manager (and now executive director), speak at a conference about how to manage cover crops without herbicides—a practical organic alternative for no-till weed control. Had he solved the puzzle of organic no-till? This is what I wanted to find out as I drove through Pennsylvania's verdant rolling hills, sweating again on a muggy July day a week after returning from Africa.

My phone guided me through the countryside where cornfields surrounded individual houses and all but enveloped a big church on a hill. Eventually, I pulled into a gravel parking lot past a white RODALE INSTITUTE sign. When I asked for Jeff at the visitors' center, I was told, "He's in the main house—go past the barn and the old stone building to the white house."

Driving away, I noticed a cloud of butterflies flitting among the vibrant wildflowers that were thriving between the fields. As I passed a great stone barn with big red doors, a guy in a white cowboy hat drove by on a tractor hauling compost. I'd arrived at the home of the longest ongoing field trial of organic, no-till farming.

Cast beneath big cotton-ball clouds, it was a nostalgically pastoral scene—a collection of white buildings with green trim and a pair of big red barns. But there were a few modern touches, like the electric golf carts shuttling people around. Water barrels collected runoff from the roof downspouts, vegetable and flower beds spread out between buildings, and a couple of geodesic domes and hoop-houses full of tomatoes stood out by the fields. A few tourists, half of them kids, were walking around the grounds, vastly outnumbered by the bees, butterflies, and insects dancing around the flowers.

I found Moyer out behind one of the barns, up on a red tractor. He was wearing a Rodale Institute hat, a red-and-white checkerboard shirt, jeans, and a major-league handlebar mustache. He's short and solid, with hands that look accustomed to dirt, and he doesn't look a day over fifty. Later, I learned he was just about to celebrate his sixtieth birthday. He speaks fast and clear, with the conviction of a man who has seen what he's talking about.

Over his shoulder by the barn door, I caught a glimpse of what I'd come to see: his tool for managing weeds for organic no-till farming. It was a heavy metal roller that looked like a steamroller drum, just not as tall. Raised, square-profiled iron chevrons protruding from the surface of the drum made it look like a stray gear salvaged from some enormous machine. But it's actually pretty simple. Mounted on the front of a tractor, it knocks down, crushes, and kills cover crops, turning them into weed-smothering mulch by pinning their stems between the chevrons and the ground, then breaking them like crimping a straw. This kills the plants and lays them down as mulch all at once. Put one on the front of a planter and a farmer can, in a single pass, plant seeds directly into fresh mulch and achieve effective weed control.

Moyer's roller-crimper makes turning cover crops into mulch simple and easy for large farms.[22] You just need the right piece of iron mounted on the front of a tractor. Rolling over the remains of the prior crop as you plant the next one offers practical, chemical-free weed control—exactly what's needed for organic no-till. By using the roller-crimper, conventional farmers can control weeds and save the money they've been spending on herbicides. Studies in other regions report that roller-crimpers provide weed control comparable to herbicides—with no resistance problem.

As Moyer led me to his office, he said that J. I. Rodale's views profoundly shaped his beliefs about the necessity of healthy soil to grow healthy food and healthy people. Once inside, I noticed a coat tree festooned with name badges from dozens of conferences he's spoken at. His opinion is in demand. After all, the institute was set up to identify problems, promote education about organic farming, and work out solutions like his roller-crimper. Over all his years at Rodale, figuring out how to farm organically without plowing has presented the biggest dilemma—the problem Moyer most wanted to solve.

Were his fields left to nature's devices, annual weeds would show up first, followed by perennial weeds, and then trees. Eventually, the landscape would try to grow back into a hundred-foot-tall forest. Unfortunately, we don't eat trees. And while you can mulch out annual weeds, you can't hold back nature forever with mulch. Jeff called perennial weeds an organic farmer's biggest nightmare.

Conventional farmers can plow or use herbicides, but Moyer compares tillage to the RESET button on a computer—it works well once in a while, but you can't be depending on it day in and day out. Moyer mostly relies on cover crops to suppress weeds because he won't spray herbicides, and doesn't like the disturbance and erosion that come with routine tillage. But while his roller-crimper can hold perennial weeds at bay for a few years, they continue to be enough of a problem that he ends up tilling about every third year. So organic no-till is a bit of a misnomer for his method. He calls what he does "rotational tillage."

Moyer's goal is to "keep something green and growing in every field all the time." He sees thick, dense stands of cover crops as the best way to combat weeds, and the right cover crops provide the added benefit of building up soil nitrogen during periods when cash crops are not growing. How much nitrogen can organic methods add to the soil? He finds that hairy vetch, a winter-hardy annual legume, can add more than 250 pounds of nitrogen per acre—more than enough to fertilize a subsequent cash crop.

Surprisingly, it's the economics of organic no-till that make Moyer optimistic about the potential for wider adoption of his methods. Consider, he said, a direct side-by-side comparison of organic corn grown under a conventional plow-based system and his no-till system. In a field previously planted with hairy vetch, growing no-till corn took a total of just two passes of diesel-fed machinery: one to simultaneously roller-crimp the cover crop and plant the corn, and another at harvest time. Meanwhile, growing it conventionally without a cover crop involved multiple passes across the field to plow, disk, pack, plant, rotary-hoe, cultivate, and finally harvest. The conventional plot produced 143 bushels per acre; the no-till one produced 160 bushels per acre. In short, the cover-cropped organic plot took less diesel and produced a greater harvest. A neighbor's comment succinctly captured the advantage of organic no-till compared to his conventionally grown crop: "I got 150 bushel corn too, and you didn't do anything, you just planted it."

Moyer laughed as he related how an economist at Penn State once responded to this example with, "Yeah, but can you prove you saved money?" All it took was a little math: with organic corn at $8.36 a bushel and conventional corn at $4.15 a bushel, the organic no-till produced a net $578 per acre profit, whereas the conventional till produced a net *loss* of $16 per acre. With organic no-till it cost less to make more.

To illustrate how well his method worked for weed control, Moyer told a story about an organic farm in Wisconsin. There the soybean harvest was 24 bushels an acre with conventional tillage, and a third more, 32 bushels an acre, for no-till involving just rolling and planting. On a field-day tour of this farm, Moyer offered to pay visitors a dollar for every weed they could find in the organic no-till field. The visitors looked all day, but in the end he didn't have to pay anyone—there were simply none to be found.

Moyer didn't seek out a job in organic farming for political or ideological reasons. He grew up on a small farm near the institute, and he went to high school with one of the Rodales. When he graduated from a forestry program in 1975 and was offered a job in Colorado, his then-girlfriend—now wife of almost forty years—didn't want to move. So he answered an ad in the paper for a greenhouse technician. The next thing he knew, he'd put in four decades at Rodale.

He didn't know it when he started, but his family had history with the farm that would become the institute, going back to when it was first cleared in the 1720s. In researching their family tree, his sister-in-law found that Moyer's great-grandmother nine generations back had been born on the farm. When Bob Rodale bought it in 1971, locals said the "organic people" would soon turn the farm into a weedy mess. Moyer didn't let that happen.

Attitudes have changed though. A couple of weeks ago, a local farmer told him, "I have to say thank you, Jeff."

Surprised, Moyer asked, "What for?"

"I came to hear you speak and thought, He's not any smarter than I am, so if he can do it, I can do it. At the time, we were thinking of getting out of farming. But we converted, and last week we bought our second organic farm, for our son. I'd never wanted our kids to be in agriculture, because it was too hard, and you couldn't make a living at it. But now that we're making money again, I can work with my son for the rest of my life."

Moyer said that, for him, this captures the mission of the Rodale Institute. "The point is to be a catalyst for change, to inspire people to change and provide them the tools. One acre at a time, one farmer at a time." He argues that modern farming could be transformed if society were to recognize that, if done properly, even intensive agriculture can improve soil health.

He rejects the idea that organic farming necessarily leads to a loss in yield, or risks starving people in a more populated future. And he chafes when conventional agronomists tend to paint organic farming as primitive. He knows organic farmers don't shun science and agricultural engineering, they emphasize practices rooted in soil biology. And these, Moyer says, are "not high-tech and human-designed, so some people think it's silly and not modern, but what we're doing is science."

Chemistry is clean and predictable—you get the same reaction every time. But biology is messy because so many things influence it. Extension agents keep telling Moyer that if he had a simple recipe, organic would be more easily accepted. But he thinks that'd be a recipe for disaster. Biology is just too variable, so the key is adaptablity. That organic farming works at scale shows "that we can grow all the food we need and still have a healthy environment." He knows this works because he's done it.

Moyer describes himself as someone who has "never been known to take the easy road." He is definitely persistent. You have to be to run an experiment for thirty-five years.

The Rodale Institute's Farm Systems Trial is America's longest-running side-by-side comparison of organic and conventional farming. It began in 1981, after a USDA study on organic agriculture identified the lack of such studies as a barrier to farmers transitioning from conventional to organic production. So the institute started a direct field-scale side-by-side test of three different farming systems: (1) organic with livestock manure; (2) organic with legume cover crops; and (3) conventional chemically fertilized. After 2008, a no-till version of each of these three systems was added to the experiment by splitting the area in each trial system. The organic no-till plots were then tilled every three years (to beat back perennial weeds) and GMO seeds were used on the conventional plots.

On the way out to the plots, Moyer introduced me to the new project leader on the field trial, Dr. Emmanuel Omondi. A native Kenyan, Omondi had been at Rodale for a month, after researching conservation agriculture for a decade in Wyoming. The three of us hopped into a golf cart–like vehicle called an E-Z-GO and puttered out to the field, where a six-foot-tall billboard-like set of panels tells the story of the field trials. Omondi walked me through it.

As a light rain began falling, he explained that if you want to control weeds, you mulch. I told him that's exactly what Anne does in our garden. He replied, "If I told you to put mulch on a thousand acres, you couldn't do it. But if you could *grow* the mulch on your field, that would be a different story." This is exactly what cover crops and a roller-crimper do.

The total area in the field trial was about the size of six city blocks (6 hectares or about 15 acres), consisting of row after row of side-by-side 60-foot-wide-by-300-foot-long plots stretching across a gentle hillside, each plot subdivided into three 20-foot-wide ones. Before the trial, the site had been farmed with conventional corn for more than twenty-five years. Rodale's study was set up with an external scientific advisory board, and results have been published in dozens of peer-reviewed journal articles, mostly written by independent, non-Rodale scientists. In addition to measuring crop yields, the standard measure of agronomic success, the study also tracked economic returns, energy consumption, greenhouse gas emissions, and soil health.

The organic systems consistently performed better by all measures, except for yield, which was comparable after an initial several-year period of lower corn yields in the organic plots. Averaged over the full duration of the trial, *including* the transition years when organic plots went "cold turkey" off chemicals, there were no statistical differences between organic and conventional yields.

The difference was only in the first three years. Without any fertilizer, after a quarter-century of conventional monoculture corn, the yields were initially lower in the organic systems. But there was no initial loss in yield on plots where the rotation started with soybeans. Lesson learned: If the soil is degraded, don't start the transition by planting nitrogen-hungry corn. Instead, begin with a nitrogen-fixing crop, like soybeans. Moyer suggested that another source of the initial yield drop was the experience gap on the part of farmers needing to learn new skills around working with rotations and new cover crops.

Right from the start, soybean yields were highest in the organic-manure system. And after the initial transition period, overall crop yields were statistically similar across all three systems, except in drought years when the organic systems produced 30 percent higher yields. The bottom line was that, after a several-year transition, there was no yield penalty for adopting organic methods. There was even a yield bonus in years with little rainfall.

Big differences in soil conditions developed, though, as soil quality and health improved in the organic plots. At the start of the field trial, in 1981, soil tests showed similar carbon and nitrogen values across all plots. By 1994, soil carbon and nitrogen concentrations had increased significantly in the organic systems but had not changed in the conventional ones. And the soil quality in the organic plots that received manure or legumes kept improving. Some plots rose from about 1 percent soil organic matter at the start of the trial to more than 6 percent today. But all the organic plots now have soil organic matter contents greater than 4 percent, whereas the conventional ones remain less than 2 percent. By conservative estimation, the amount of carbon in the soil doubled over several decades. A number of other studies have also reached the same conclusion: that organic farming and organic-matter input increase soil carbon and nitrogen content as well as microbial biomass and activity.

This matters. By increasing soil organic matter, aggregate stability is improved, which, in turn, allows more water to infiltrate into the soil instead of running off.

To see these changes, all you need is a shovel. When Moyer and I marched off into the plots to take a look, we found that soil from the organic plot was brown, moist, and composed of stable aggregates perforated by visible pores. The conventional soil had few visible pores and was yellower, drier, and disaggregated if held. This means that the conventional soil falls apart easily and sheds erosive runoff. Here, Moyer says, is why Iowa is washing down the Mississippi.

In terms of energy use, the organic systems used just under half as much to produce the same harvest. The difference mostly lies in the energy that's needed to produce fertilizer and herbicides for the conventional plots. All in all, the organic systems produced comparable harvests while increasing soil organic matter, improving soil health, and using less energy. That sounds more efficient to me.

I found the economic analysis more surprising. Conventional systems produced higher profits than organic systems for the first several years when they outpaced the latter in yields—if you assumed that both organic and conventional crops receive the same market price. After that initial period, the two systems were comparable. But organic and conventional crops generally don't receive the same market price. When the price premium for organic produce was factored in, organic production was over three times more profitable than the conventional system. The lesson was that if conventional farmers switched to organic they could maintain high yields, improve their soil, and make more money.

Moyer describes the Rodale Farming Systems Trial as pulling back the curtain on the feasibility of transitioning conventional agriculture to organic through a rigorous scientific analysis. He emphasizes that it doesn't matter what metric you use—water, energy, soil health, profitability, or nitrate pollution—organic production performs better, except for yield, which is comparable after the transition. This is the point that high-profile, meta-analysis comparisons of conventional and organic farming can miss. While there has been very little long-term research on organic production, Rodale's thirty-five-year field trial has shown that intensive organic farming offers an economically viable, energy-efficient alternative for maintaining high crop yields. Other studies also suggest that the organic yield penalty is greatly reduced, if not eliminated, after the several-year transition period.

This raises the question: Why is the USDA not going all out to support research on and a transition to agricultural practices that not only deliver yields comparable to conventional yields, but greatly lessen or solve many of the problems that plague conventional agriculture? Could it be because that's not what influential Big Ag suppliers might want? Moyer wants to see a new research imperative for the USDA to investigate how to adapt farming to regenerative practices in different regions and with different crops to achieve yields comparable to conventional, reduce erosion, and improve soil health. "You look at the data, you look at the science, and you have to come to the conclusion that we have to change."

I asked Jeff, "Why aren't more farmers converting to organic methods?" Without missing a beat, he pointed out that crop insurance zeros out the risk of bad years for conventional farmers. "There are two ways to drought-proof your soil—one is with biology and the other is with crop insurance." He likens a farmer with crop insurance to a gambler heading to Vegas with $100,000 and the guarantee that, if he loses, the casino will give him the money back and he'll get to try again. Farmers will always take big risks with a deal like that. "Conventional corn is like a racehorse," Moyer told me. "It runs great on a perfect track. Organic corn is like an old workhorse. It produces reliably but won't win big." He says organic farmers don't need much crop insurance because what they do is not as risky, since yields from organic plots are more reliable and consistent.

I got the impression that at first the USDA was suspicious of the Rodale results. Then they got curious. After a decade of observing the institute, they set up a parallel plot experiment in Maryland to test the radical contention that organic farming could perform as well as chemical farming. So did some universities—Wisconsin, Minnesota, Iowa State, and UC Davis. A 2015 review of all six long-term organic comparison trials found that organic practices proved economically viable (comparably or more profitable) and resulted in improved soil quality, greater soil carbon capture, and better pest suppression. Yield differences were more variable, depending on particular crops, and soil and weather conditions. However, at several of the sites, organic yields consistently equaled or exceeded conventional yields. And at four of the six sites, a farmer's growing experience with weed management improved organic yields over time. The greatest yields in the organic plots were achieved through combining legumes in crop rotations with periodic additions of livestock manure.

For me, the takeaway message of these studies was that cover cropping and crop rotations help build soil organic matter (carbon) and soil health in ways that promote short-term nutrient availability and long-term fertility in *both* conventional and organic farming.

But, could we really feed the world if every farm converted to organic tomorrow? A 2015 meta-analysis published in the *Proceedings of the Royal Society* offers the most extensive comparison of yields from conventional and organic farms, drawing from 115 studies with more than a thousand observations. First, the authors did the usual, simple comparison of yields for organically grown crops versus conventionally grown crops. They found that organic production averaged 19 percent less. This finding is similar to prior meta-analyses invoked by the organic-can't-possibly-feed-the-world crowd.

Yet the extensive dataset the authors used in the 2015 study allowed for an additional comparison. They further analyzed the yields of only those organic crops that incorporated cover crops and rotations versus conventional methods. They found that the yield gap was much less, more like 8 or 9 percent. The authors concluded that research to improve organic systems could further reduce or close this gap. Moreover, the gap may be smaller than it appears, as the authors found "evidence of bias in the meta-dataset toward studies reporting higher conventional yields."[23] The bottom line? There is little yield penalty when organic practices also integrate practices that build soil fertility—like cover crops and diverse rotations.

Along these lines, an eight-year (2003–2011) University of Iowa field study directly compared a conventional two-year corn-soybean rotation with more diverse three- and four-year rotations that received five to seven times less nitrogen fertilizer and six to ten times less herbicide. Grain yields were consistently higher under diverse rotations, despite greatly reduced chemical inputs. The more diverse system also proved more profitable. Like the authors of the Royal Society paper, these researchers contend that the commonly reported yield gap between organic and conventional farming could be overcome with very low inputs of agrochemicals in diversified cropping systems. The more I looked into it, the more hollow the standard line that organic farming can't feed the world sounded. It mattered *how* the organic farming was done. And organic-ish practices might provide a practical option for conventional farms.

Moyer particularly emphasizes that organic farmers need to adapt their practices to the specific conditions of their farms. It's like a ski racer who sets out to win a race. Every course is different—snow conditions, weather, moguls, steepness, and so on. Racers take all this into account to craft tactics and moves that they hope will put them at the front of the pack. The trick to organic farming is to get to know the land and, through experimentation with cover crops and rotations, find winning combinations for each region and farm.

Omondi wants to add another element to the field trial and test a hybrid system combining conventional and organic methods. He wants to experiment with the use of cover crops and legumes to minimize herbicide and fertilizer use. He thinks that when conventional farmers see they can have good yields with fewer chemicals they will flock to the cost savings. They might not go fully organic, but they could save money and greatly reduce their environmental footprint. He's thinking right in line with Dwayne Beck and Kofi Boa.

While we were standing around talking, four guys in their twenties walked out to the field. They'd come from the Hudson Valley, where they farm 20 acres of organic vegetables. They're considering going no-till and wanted to see the Rodale demonstration plots. Moyer guided them into the cornfield to explain to them what they're seeing.

He'd already given me plenty to think about, so instead of following them, I watched high-rising cumulus clouds float across the sky and thought about the blurring lines between conventional and organic practices. What was the significance of conventional farmers moving toward organic-ish practices and organic farmers reevaluating the utility of using small amounts of chemicals? I was starting to think they are both on to something.

AGRICULTURE IS IMPERATIVE

The next morning, I skipped the abominable breakfast buffet at the Holiday Inn Express and set off toward the institute, looking for something to eat as I drove through a landscape of corn and soybeans. On the way into Kutztown, I found a huge twenty-four-hour grocery store. Hunting through two full aisles stocked wall-to-wall with chips, I finally came upon the lone variety of whole-grain organic ones hidden behind a garish floor display. After wandering some more through this wilderness of processed corn and soy products, I arrived at the organic produce section. It held a few pathetic plastic-wrapped avocados and bell peppers. Organic advocates have a long way to go if they aim to change American agriculture. None of what they are doing will matter much if farmers can't get their goods distributed and marketed so shoppers can (and want) to buy them.

When I arrived at Rodale, Kristine Nichols, their new chief scientist, met me in the parking lot. She ushered me into a single-story, white-and-green-trimmed building divided into cubicles and small offices. Hers was across from one stuffed full of "Soil Health Kits" she'd put together to let gardeners and farmers assess the health of their soil.

Talking over her desk, I learned that she'd just passed her one-year anniversary at the institute and had her hands full managing more than a dozen projects and a crew of interns, technicians, and scientists.

Nichols exuded the casual confidence of a field scientist in sandals, a light blue shirt, and dark blue pants, with black-rimmed glasses and a narrow white headband holding back her shoulder-length red-blond hair. She came to Rodale from the USDA's Agricultural Research Service in Mandan, North Dakota. Our conversation quickly turned to science, and it didn't take long to learn that she's perfectly happy to dive down rabbit holes of detail I struggled to absorb and capture in my notes.

Looking up, I'm distracted by the three-foot-long AGRICULTURE IS IMPERATIVE sign that dominates her office wall. Turns out the staff gave it to her when she started, since, when asked at her interview for one word to describe agriculture, she came back with "imperative." Half-jokingly, she told me that she thinks that's what got her the job. The sign was already up in her office on her first day.

She sees soil science undergoing a major evolution in thinking as soil chemistry and physics make way for soil biology. The change is mostly driven by farmers and researchers who want something new to help solve the same old problems. "It's not that chemistry or physics are wrong," she said. "But how we've been thinking is not the full picture—something is missing."

At the root of this new line of thinking is that soil is not just created by the elements. Rather, it's a complex mix of geology and biology. Things that happen in soil involve messy, complicated biogeochemistry. Microbiology matters as much as chemistry and physics—and always has.

As you read earlier in Chapter 3, the very first land plants had mycorrhizal relationships. At first, roots were simply plant anchors and not absorptive structures, because fungi did the absorbing for plants. Lichen, bacteria, and fungi broke down minerals. Only after soils began forming did plants evolve roots that could suck up the available soluble nutrients. Nichols likens it to a chicken-and-egg puzzle as to which came first, plants or soil. "You couldn't have plants without soil, or soil without plants—the missing link was the fungus." Over a half-billion years of life on land, fungal-plant partnerships evolved remarkably efficient water and nutrient management. She says the only way to do it better would be to manually inject nutrients directly into plant roots several times a day. This means that microbiology offers a great way to regenerate fertile soil.

As noted previously, standard soil tests measure chemical nutrients that are available to plants at that time. While this information has a lot of value, it can't account for biological interactions in the soil, which influence a great number of things. Phosphorus, for example, does not stay in a soluble, plant-available form for long. While most soils have enough phosphorus to grow crops, most of it is not plant-available without microbial assistance. "It's like we're here in Kutztown and all of our phosphorous is in Philadelphia," Nichols said. Microbes run the delivery service. While few fungi have good enzymes to break down minerals, phosphate-solubilizing bacteria that colonize mycorrhizal fungi form partnerships to scavenge nutrients from minerals. With their yards and yards of hyphae, fungi absorb nutrients and transport them over long distances to plants—in exchange for carbon-rich exudates from plant roots. So, to take full advantage of biotic fertility a farmer needs soil filled with beneficial bacteria and mycorrhizal fungi.

Agricultural soils generally see big losses in mycorrhizal fungi abundance and diversity and become dominated by a few species of bacteria. It's not hard to understand why tillage, in particular, chops up hyphal networks, severing nature's finely honed nutrient delivery system. On top of this, when you look at microbial community diversity, roughly half of soil organisms are as of yet unknown, so it's hard to nail down what the other effects might be and there's a lot of controversy about what's actually happening. Still, thinking about microbes as a key to unlocking bound-up nutrients, like phosphorus, is crucial when thinking about soil fertility.

When crops get doused in fertilizers, what they don't take up gets washed out or leached out of the field and flows downstream. And, the plants don't give out as much exudate. "Why pay for something that's already free?" Nichols asked. But, as this leaves the plants without their microbial partners, heavily fertilized crops don't get a full complement of micronutrients. When conventional farmers question the utility of adding insoluble organic nitrogen in compost or manure when it's not immediately plant-available, they're overlooking the missing piece of the puzzle: the biology makes nutrients available.

I asked Nichols how she got interested in fungi. Her dad and uncles were farmers, and she wanted to do something different. As an under-grad at the University of Minnesota, she was interested in genetics, but her genetics professor did not respond to her inquiries about working in the lab. Her plant biology professor did, and so she spent three and a half years working in a lab studying mycorrhizae.

By the time she graduated, she was hooked on fungi. Nichols applied for a master's program at West Virginia University, hoping to work with an expert on fungal evolution. She was accepted, and her advisor encouraged her to think about how plants and fungi worked together. After she took a biogeochemistry course, her thesis zeroed in on help-ing USDA scientist Sara Wright study glomalin (pronounced GLOW-mell-in). This curious stuff is a protein that mycorrhizal fungi make in the walls of their hyphae and exude out into the soil. The first thing mycorrhizal fungi form is hyphae; glomalin is the second—hyphae need it to work properly. And this means that if fungi are to help build soil fertility it's essential that enough carbon is available in the soil to feed the fungi.

When Nichols was finishing up her M.S. thesis, Dr. Wright recruited her for a Ph.D. program at the University of Maryland with support from the USDA. One of her M.S. thesis committee members asked: "Why would microbes put so much energy into making a compound that just goes out into the soil?"

This was like the question of why plants push carbon-rich sugars and other molecules through their roots out into the soil. But with glomalin, there's a twist. It's hard to break down, so it doesn't feed anything. But it seems to "weather seal" the porous walls of fungal hyphae, which are otherwise like pipes full of holes. The glomalin acts like a polymer coating, sealing leaks where necessary. This allows hyphae to transport material over long distances in the soil across pressure changes in pockets of air and water.

Glomalin also helps aggregate the soil. It's sticky, glue-like qualities bind small particles together. And its wax-like property that seals up hyphae makes some soil pores impermeable to water, but not to air. So the air in pore spaces can escape when the spaces fill with water in a soil with aggregates stabilized by glomalin. But when water fills pores in a soil with aggregates that are not stabilized by glomalin, the air can't escape, which can increase the air pressure in the remaining voids enough to break up the aggregates. Thus, fungi make the glue that binds microaggregates together and stabilizes passages through which water moves and can be stored. These spaces are essential for the fungi to thrive and also serve as habitat for soil life. In other words, the physical structure of fertile soil depends on its biology. Physics, chemistry, *and* biology are at work.

This is what conventional agronomists missed. A conventionally tilled system may produce soil aggregates, but they won't be stabilized by glomalin. In contrast, no-till farming with cover crops provides ample carbon for mycorrhizae to consume and convert to glomalin. This means that life-rich soil promotes infiltration and holds on to water better, so plants have a greater backup supply through the dry season. This also helps to explain why tilled soils disaggregate so readily, shed more runoff, and don't hold water or fertilizer as well as undisturbed soils.

Stable soil aggregates are also important for regenerating soil organic matter. When organic matter is held within a stable aggregate, its turnover time increases from years to decades or longer. That's part of why soil rich in mycorrhizal fungi can sequester carbon more rapidly than lifeless dirt can. It's another reason why fungi that make glomalin would themselves prove so successful—it makes the soil more fertile and that, in turn, produces more organic matter for the fungi to consume and break down.

This, Nichols says, gets to the heart of the big problem in agriculture. If we simply managed plants to build soil carbon and promote soil health, we would fix everything else. "We spend all our time putting Band-Aids on this issue. Our farm subsidy programs should not to be based on yield—they need to be based on soil health. The problem is that we're paying money to people to farm stupidly and destroy the infiltration capacity of the soil." She nodded vigorously when I suggested that, with government subsidized crop insurance, we've socialized risks and privatized profits.

If people paid more attention to the condition of their soil, things would change. Nichols suggested that "the most important tool in agriculture now is a shovel—dig a hole and look at your soil." With that, her eyes lit up mischievously. "Let's go see a soil pit." She had a trench opened up across a couple of the plots in the farm systems trial, and she wanted to show me this window down into the soil.

We bumped back out to the field-trial plots on an E-Z-GO with a pickup-truck-like minibed on the back. We stopped and walked over to a corner of the fields, where a three-foot-deep backhoe pit had a ramp slanting down on one end for easy access. The pit cut across two plot treatments—conventional and manured organic—that have been running since 1981. Both were converted to no-till in 2008.

The floor of the pit was comprised of weathered shale bedrock, which overnight showers had left a bit sticky and muddy. In one of the long walls of the pit, the soil beneath the conventional no-till plot was exposed to the left, and the soil beneath the organic no-till plot was on the right. In the face of the pit wall, the bedrock graded indistinctly upward into yellow-tan subsoil, then an abrupt transition to darker brown topsoil and matted organic matter at the surface.

Nichols pulled some white flags mounted on the ends of thin white rods from the bed of the E-Z-GO. Together we bent down to inspect the soil profile, noting the distinct transition from topsoil to subsoil. We stuck the flags into the trench face to mark the base of the topsoil, then measured down from the surface to assess the thickness of what soil scientists call the "A horizon," the zone of organic and mineral matter most of us know as topsoil. In the conventional plot it extended nine inches down, while the organic plot's A horizon was twelve inches. They'd started out the same, so the organic plot's manure-organic system had built an extra three inches of topsoil since 1981. This comes close to an inch of soil per decade. Here was as direct a demonstration as one could hope for in terms of showing how intensive production agriculture can build soil.

This is huge. By building soil from the top down, the Rodale farmers were reversing the ancient story of agricultural soil degradation and making soil health a consequence instead of a casualty of agriculture.

Sailing back to the office on the E-Z-GO, I asked Nichols why she'd left the USDA. She told me that she'd felt frustrated that the Agricultural Research Service had strayed from what she thought its primary mission should be: helping farmers. They were not doing farm trials, plant breeding, or plant pathology. Instead, there was constant pressure to chase after funding and hot topics—"that's what academics do!" To her dismay, agency researchers were *not* supposed to talk to actual farmers. She was supposed to give her results to extension agents, and *they* would talk to farmers. She'd been told, "Your job is research, so do your job." She felt like the agency was on a trajectory to becoming irrelevant to farmers if peer-reviewed publications served as the only way to measure success.

Still, she said that the best education she had was during her time with the agency in North Dakota. It got her thinking about farming systems. Every year the Red River would flood, and year after year farmers would till up the soil to grow potatoes and root crops. In the process, they would destroy the soil structure, so rain and snowmelt would run off instead of sinking into the ground, which caused erosion and exacerbated the flooding. But when their fields flooded, farmers would get money to replant and do the same thing, over and over again. The whole system made no sense.

What made far more sense was what she saw no-till farmers doing. One in particular she mentioned was Gabe Brown, who hadn't qualified for crop insurance because he planted cover crops. The thinking behind this policy was that because green plants take up water, growing cover crops would reduce yields. It didn't matter, apparently, that it didn't actually work out that way. When Nichols visited Brown's farm, he asked her why he was outproducing his neighbors when what he was doing went against conventional wisdom. She didn't have a ready answer for him.

And this takes us back to the old question of alternatives to the plow for weed control, and the value of cover crops and crop rotations—the proverbial pillars supporting conservation agriculture. Perhaps, I thought, we should just get on with applying these simple principles on all our land and stop arguing so much about conventional versus organic. Innovative organic farmers and conventional no-tillers were coming to similar conclusions from different directions. If both camps are converging on a similar set of practices, then why not push for converting all of agriculture to these soil-health-promoting practices? If two culturally distinct, reflexively oppositional groups come to embrace the same principles, it's probably because they work!

This presents a revolutionary prospect—to move conventional farming to regenerative practices. I particularly liked how Jeff Moyer put it: "If you're conserving, you're holding on to what you've got. With regenerative agriculture, we can have more if we use it right." I also like how regenerative connotes thriving. For if you regenerate the soil and regenerate farms, you regenerate farming communities. And since it only takes a few years to transition to profitable organic or organic-ish, low-input farming, change can happen fast.

But Nichols wonders if we are still missing a piece. Based on what she saw on Gabe Brown's farm in North Dakota, she suspects that it may also be necessary to reintegrate livestock into cropping. So Rodale is starting to experiment with a version of double cropping, in which cows forage on crop stubble. This way, the cows eat the grass part of the corn that they'd evolved to consume, we eat the grain part—the corn—and the soil gets a steady diet of cover crop residue and manure. Of course, we also get to eat the cows, or the milk (and cheese) they produce.

But I suspect that a bigger part of what was motivating Nichols' interest in animal husbandry was how livestock grazing makes plants produce a different set of root exudates than those they produce if neatly cut or mowed down. All the pulling and tugging from the grazing action pulls out root hairs, causes ragged tears, and creates leaks. Grazing creates a lot of plant wounding, and the plants must scrounge up more resources from the soil to patch themselves up. So they push out more carbon-rich exudates into the soil to recruit microbial assistance to help repair the damage. In this way, grazing may accelerate soil building because those carbon-rich exudates prime the pump of soil life and build up soil organic matter.

Rodale's flirtation with grazing was catalyzed by a neighboring dairy farmer. On his 40-acre farm, this gentleman kept his sixty-cow herd in the barn and fed them with corn. When the price of corn went up and the price of milk went down, he went to his bank and said he had to sell the cows. But the bank told him no, since it owned the cows. So Rodale bailed him out by aiding his transition to organic and by grazing his cows on their newly planted pastures—a win-win situation.

Friendly know-it-alls told the dairy farmer that his cows wouldn't know what to do when he let them out to graze. Yet when he opened up the barn door, they immediately took off running into the fields, "like schoolgirls running out for recess." Now he no longer rents the farm but has purchased it. And each year Rodale collects grazing fees as the cows' chief by-product—manure—helps build soil carbon and improve soil health.

As we headed back to my car, I thought about how Rodale's long-running side-by-side comparison shows that organic no-till practices can sustain crop yields, build soil, and put more cash in farmer's pockets, and do it on a full-farm scale. I got into my red rental sedan and rolled down the window just as the wind shifted direction. The smell of composting manure hit my nose. It smelled sweetly scorched, bitingly organic, but not truly offensive—the smell of life returning to the land. How much of a difference could bringing livestock back onto farms really make? More than I imagined as I headed off to visit Gabe Brown in Bismarck, North Dakota.

9

CARBON COWBOYS

All is not butter that comes from the cow.

—Yiddish proverb

Most environmentalists—and environmental scientists—consider cattle a blight on the land. It's no secret that overgrazing can strip away vegetation and promote erosion of bare soil. Until recently, I too might have scoffed at the idea that intensive grazing could heal a landscape and restore soil fertility. Until, that is, I saw what Gabe Brown had done on his North Dakota ranch using cows to rebuild soil health.

I didn't come around to this view easily. My Ph.D. research had focused on understanding where stream channels begin, and one of my field research sites was in the grasslands of Tennessee Valley, just north of San Francisco. Walking the watershed to map channel heads, I couldn't help but notice the deep gullies carved into the valley bottoms.

One day, the glint of California sun reflecting off snow-white cow bones exposed several feet down in a gully wall caught my eye. Intrigued, I began scouring the walls for pieces of charcoal or wood to radiocarbon date. A pair of samples from around nine feet down came out at about 6,000 years old. The deepest sample I found, exposed in a gully wall about fifteen feet below the valley bottom, was more than 9,000 years old.

The story came together. After the end of the last ice age, Tennessee Valley slowly filled in with sediment. Until, that is, cows arrived in the mid-nineteenth century and gullies cut back down to bedrock. In fact, archaeological surveys conducted before 1907 report the existence of a native Coast Miwok site exposed in an incised gully wall. And, in the late 1980s, an octogenarian who grew up in the valley told me how, as a young boy, he saw men plant eucalyptus trees—now huge—to stabilize collapsing gully walls. So why did the valley bottoms fall apart in the late 1800s?

We know the Tennessee Valley area was used for occasional hunting before it was leased for dairy cattle grazing from the 1850s to the 1890s. County tax records show that during this period the number of cattle quadrupled. By the close of the century, northern California's growing gullies caught the attention of a civil engineer who attributed them to overgrazing "so close and continuous, that the forage plants and grasses have nearly disappeared."[24] With no climate shift or other factor to point to, I too concluded that overgrazing destabilized the valley bottoms.

This experience and others like it reinforced my opinion, and I was certainly not alone in thinking that intensive grazing was a surefire way to unravel a landscape. Then I met Gabe Brown.

In January 2013, he and I both spoke at a conference in Salina, Kansas. His description of how he improved his farm's soil impressed me. What really surprised me though was his emphasis on intensive grazing as the key to soil building. After converting his 5,000-acre ranch to mixed livestock, no-till, and rotations with cover crops, he uses far less agrochemical inputs *and* outproduces his conventional neighbors. When I started visiting farms around the world, his was high on my must-see list, even before Kristin Nichols from Rodale mentioned his name.

When we first met, he also told me that he was an advocate of smaller farms, because he sees farmers' ability to make a good living on a small farm as the key to revitalizing small-town America. I think he's right, even if the idea of smaller farms goes against the grain of recent history. In the second half of the twentieth century, a bigger-is-better philosophy shaped agricultural policies and subsidies that promoted mono-cultures and divorced animal husbandry from crop production. As the size of the average American farm tripled between 1930 and 2000, from about 150 to 450 acres, the foundation of farm income shifted from diversity to specialization. Farmers became very good at growing a few crops. This increased commodity crop production, which raised supply relative to demand and drove crop prices down. But at the same time, costs rose for the chemical inputs that were becoming increasingly central to agricultural production.

Small farms began disappearing as farmers got squeezed in a system that prioritized commodity production over farm profitability. In 1930, one U.S. farmer fed a dozen people. By 1990, an American farmer fed a hundred of their fellow citizens. Larger farms translated into fewer people on the land and ultimately sapped economic vitality from small towns across America.

Brown thinks that rebuilding fertile soil is the key to making small farms profitable again. The trick is to maintain yields while slashing input costs. That's what he did. But Brown would be the first to say that getting this done isn't easy, especially if you're stubborn and independent and wary of change, like many farmers. He forgoes government subsidies and didn't go organic. And why pay someone to certify that what he does is better than what conventional farmers do, when he thinks anyone can see that it is? He bristles at anyone trying to limit his choice of tools, even those he doesn't use very much or often anymore—like fertilizers and herbicides.

Clad in a baseball hat and jeans, Brown didn't look like an award-winning prophet of sustainable agriculture when he pulled up in his gray Dodge Ram 1500 at my hotel on the edge of Bismarck, North Dakota. Short, clean-shaven, and balding, he's plainspoken, likes to talk with his hands, and employs a quick, self-deprecating wit. "Now, I'm not that smart," he'll start off before delivering an insightful and concise indictment of conventional agriculture.

We drove out to his ranch, past cookie-cutter houses encroaching on farmland and spilling into open prairie. But before we got to his farm, Brown told me he wanted to show me the soil on neighboring farms.

Across from his long driveway, we pulled off the gravel road and parked on the grass in front of a downed NO TRESPASSING sign. Brown led the way into his neighbor's dense stand of three-foot-high sunflowers. Since the late 1960s, this field has grown a low-diversity, two-crop rotation of mostly flax and wheat. The sunflowers there now were only planted to salvage a crop after more than three inches of spring rain fell in less than an hour and eroded off the original wheat crop. Although it's been no-tilled for years, the field has always received a lot of fertilizer. And there's hardly any crop residue on the ground, nothing to hold the soil in place.

Brown thrust a shovel into my hands. I set the blade and stepped on the top lip, sending it into the light-brown, silty soil. It was hard, dry, and difficult to break up—as he knew it would be. He told me that it tests out at less than 2 percent organic matter. That looked about right to me. There was no trace of soil life—nary a worm to be seen. Inspecting one of the seedheads, Brown shook his head. "Seed weevils are moving in. They'll be spraying this soon."

The second field, farther up the road, is maintained by three brothers, a no-till farm of 40,000 mostly rented cropland acres. Considered top farmers by their peers, they follow a corn-sunflower-barley rotation and some years use over 400 pounds of nitrogen per acre. Brown says that the high chemical inputs have killed off their mycorrhizal fungi, so despite going no-till, they haven't improved their soil all that much. When I dug down, their silty brown soil was dry and platy, much like the first field. With no binding aggregates to hold it together, this fragile dirt crumbled into dust when squeezed.

Our third stop was a neighbor's field that's been farmed organically for more than a decade. The first thing I noticed was all the bees. Unlike the silent fields we'd just visited, this one hosted an insect serenade. And though this soil was a bit darker brown than the other two, there was not much texture; it was blocky, without aggregates. Brown told me that water does not infiltrate into this field, despite its history of a diverse rotation that includes flax, wheat, soy, alfalfa, clover, sunflower, corn, oats, and barley. Precipitation tends to pond and run off, because "the soil is capped." This field has been tilled habitually, and tillage breaks up mycorrhizal fungi, which means that there won't be much, if any, glomalin to hold soil aggregates together. With no aggregates to create soil structure and pore spaces, it won't infiltrate and hold water. This soil was free of chemicals, but its physical structure was far from ideal for growing crops.

All three of these farms had the same type of soil as Brown's farm—the Williams loam, the most common soil in the area. So differences are attributable to farming practices alone. While his neighbors used some of the principles of conservation agriculture, none used the full system. Brown stressed that you need to adopt all three—no-till, cover crops, and a diverse rotation—to get better-than-conventional results and build up soil fertility. It also helps, he said, to use a fourth tool—livestock.

"Let's go see if you can see the difference in our soil," he said to me as we walked back to the truck. I couldn't help but notice the contrast between what I'd just seen and Brown's land across the road, even from a distance. The frontage to his ranch wasn't grass; rather, it was a strip of native wildflowers he'd planted to support pollinators. He partners with a beekeeper to host hives and harvest honey. When the bees finish with the flowers, his cattle graze on them. The cows aren't picky and they convert flowers into meat, something Brown can then turn into dollars. I suspect that such thinking is a big part of why he's doing so well.

Driving the long gravel driveway that leads into his farm, we passed lines of newly planted trees—nuts on the right, fruits on the left. After passing several outbuildings, green farm equipment, and his modest ranch-style house, we headed down to a field where hairy vetch climbed corn stalks and vegetables grew between the rows.

Cacophonous insects greeted us as we walked into the field. Closing my eyes, I could imagine I was back doing fieldwork in the tropics. This time, when Brown handed me the shovel, it slid down effortlessly into the silty soil. I picked up a handful and found it moist, dark brown, crumbly, and shot through with macropores (big holes). It was capped with a half-inch blanket of crop residue. Brown's soil bore no resemblance to his neighbor's crusty dirt.

When he bought the place in 1991, his soil had less than 2 percent organic matter, just like his neighbors'. By 2013, he'd more than *tripled* it. Brown had managed to restore his soil—and its fertility—through profitable intensive agricultural production in a couple of decades, the blink of a geological eye.

While it took a while to figure out how to build soil, he says he could do it much faster now. The key is to cultivate the mycorrhizal fungi that promote glomalin production, build soil aggregates, and increase soil organic matter in conjunction with carbon-rich root exudates. This means adding organic matter through cover cropping and not disturbing the soil through tillage. But he says the real secret to soil building is intensive grazing. And rebuilding his soil paid off in ways that surprised him. That spring storm that ruined the neighbor's wheat didn't damage Brown's fields at all. If anything, it increased his harvest.

A BETTER WAY

Back in 1993, the Natural Resources Conservation Service (NRCS) came out to assess Brown's fields. They measured that it took an hour for half an inch of water to seep into the ground. At that point Brown was just starting the process of rebuilding his soil. In 2009, he got to see the fruits of his labor. That June, an intense storm cell dropped about thirteen inches of rain on Brown's farm in under six hours. The next day his soil was wet, but he didn't have any standing water, unlike his neighbor, who had lost his crop to erosion and the rainwater ponds that had sprouted in his fields. Two days after the storm, Brown could have driven just about anything across his own fields without rutting them up. Why the difference? On Brown's fields all that rain had infiltrated into the ground.

When the NRCS came back and measured his fields again in 2011, eight inches could infiltrate in an hour. And by 2015 it took less than half a minute for two inches of water to sink in. More water infiltrating instead of running off translates into higher soil moisture and greater drought resilience.

Like many modern American farms, Brown Ranch is split into blocks scattered across the landscape. Brown owns almost 1,400 acres, but works a total of about 5,000 acres of low rolling hills with no more than 100 feet of relief to the horizon. His farm is smaller than average for his area, but he's not trying to get bigger. He wants to shrink his farm and only work land he owns.

Brown grew up in Bismarck, several generations after his family settled in North Dakota in the 1880s. In 1991, he bought the home section of his in-laws' farm. They had conventionally farmed it since 1956, with regular tillage, lots of fertilizer, and a small herd of cattle. The first year after Brown took over, good harvests still depended on a lot of fertilizer and pesticides. The next year, a friend advised him to go no-till to save time and effort. He thought that made sense as a way to save water, a huge concern since his land normally receives an average of just sixteen inches of precipitation a year—ten from rain and six from snow.

When Brown went no-till in 1993, his father-in-law was embarrassed and perplexed, convinced that "the more you worked the soil the better." But as things went surprisingly well for the first few years, Brown diversified with hairy vetch and peas for livestock forage.

Then the day before harvest in 1995, Brown lost his entire wheat crop to hail. He barely managed to plant a quick cover crop for the cattle to graze. What do you do with hailed-out wheatland, especially if you're a bit stubborn? Plant wheat again, like Brown did in 1996. Then hail returned for a second round. Brown had to take an off-farm job to pay the bills.

The next year, 1997, was very dry. He didn't harvest a thing, and nobody else around did either. Only a loan from his family kept the bank from foreclosing. On top of that, he lost 80 percent of his next crop to—you guessed it—hail, in 1998.

This was a disastrous run. But looking back, Brown says that four years in a row of failed crops was "absolutely the best thing that could have happened." Those four years—and his resulting desperation—pushed him to change the way he farmed. He couldn't afford any inputs, so he had to figure out how to grow crops without them.

At the start of a long reeducation, Brown, curious about the use of cover crops, delved into Thomas Jefferson's farm journal. It made him want to trade all of his agronomy books for ecology ones instead. He realized that modern agriculture had taken diversity out of the equation after he read *Buffalo Bird Woman's Garden*, a 1917 account of a Hidatsa woman who grew a wide variety of corn and beans in alternate rows along with squash and sunflowers. Her story illustrated that rotating crops isn't really anything new. Brown's methods simply update old ideas that work.

Soon, Brown started noticing subtle changes in the soil. It smelled different, retained more moisture, and looked darker. Back then he hadn't ever heard the words "soil" and "health" strung together, and he thought of a cover crop as something to use to keep erosion from happening. But it was becoming apparent that planting cover crops and not tilling improved his soil. And he liked that.

By 2000 he could afford to use fertilizers again. So he did. A few years later, in 2003, Kristine Nichols (who worked for the Natural Resources Conservation Service in North Dakota at the time), asked him why he was back to using synthetic fertilizer. Why not use soil biology instead to cycle nutrients? Intrigued by her questions—and attracted to the potential to save big using less fertilizer—he conducted side-by-side field trials from 2003 to 2007. Much to his surprise, he got equal or greater yields with no fertilizer.

Now he had some questions of his own. Had he already improved his soil enough that fertilizers wouldn't help? Had he been writing big checks for nothing? Since then, Brown hasn't added any synthetic nutrients to his fields. Instead, he relies on legumes to provide nitrogen to the grass, and grasses, through mycorrhyzal fungi, to provide phosphorus to the legumes. He hasn't used pesticides or fungicides since 2000 and quit using pesticide-treated seeds a few years later. Finding he could maintain yields without paying for a lot of chemicals was a welcome discovery: "I like signing my name on the back of the check, not the front!"

Today, his goal is to keep advancing soil health and eliminate herbicides altogether. He still uses them occasionally, about every third or fourth year to beat back perennial weeds like Canada thistle. This is about the same interval that Jeff Moyer tills to manage perennial weeds at the Rodale Institute.

For Brown, the next piece of the soil health puzzle fell into place when he heard no-till champion Ademir Calegari speak about planting cover crops in multispecies combinations. Two things the Brazilian agronomist said resonated with him. Cover crops need as little as two inches of rain to grow and they were meant to be planted in mixes of more than half a dozen species. So Brown tried a multispecies cover-crop mix in what turned out to be a drought year. His crops did far better than expected; so much better, in fact, that he bowed out of government crop insurance. A decade later he doesn't regret it. "I don't want to be on welfare. Why should the taxpayers have to subsidize me?" Brown says he's not only doing better than those who stayed in the program, he's building resilience back into his land.

I couldn't help but appreciate the way he frames the effect his new practices have had on his perspective. "When I was farming conventionally," he said, "I'd wake up and decide what I was going to kill today. Now I wake up and decide what I am going to help live."

Shifting into a solemn tone, he told me that many of those in the regenerative agriculture movement are Christians seeking to be good stewards of the land that they'll someday pass on to their children and grandchildren. His faith is central to his enthusiasm about restoring his soil. It's about taking care of Creation. But he also knows that if it's to last, it has to pencil out as well. Fortunately, a regenerative model is profitable.

Paul, Brown's twenty-seven-year-old son, is a taller, thinner version of his father. He's also part of a new generation of farmers who have never tilled the soil. And he happens to be business-savvy; he pushed his folks to get into direct-to-consumer sales and to start their own trademarked "Nourished by Nature" brand to sell. They've been pleasantly surprised by the demand, and how consumers are starting to view nutrient-dense food grown in healthy soil as preventative medicine essential to their own health.

Brown wanted to make sure I knew that you don't have to tie up a lot of money to do regenerative farming. You can do it at fairly low cost. Owning a lot of high-value depreciable assets, like combines and big sprayers, is what got many farmers into trouble in the 1980s. A conventional farm can go upside down remarkably fast, from $1 million net income one year to $1 million loss the next. All it takes is one bad storm or a too-dry summer. That sounds more like Wall Street roulette than a stable family business. Brown is certain that the financial diversity of his system is better for people and the land.

So what's the strategy for growing small farms? Brown points to how turning farm waste into profits emulates nature's efficiency and makes good business sense. Stacking enterprises in this way produces better margins and a more resilient, diversified income stream. At the heart of it all is rebuilding fertile soil, although he's quick to add that it would help to reform or eliminate conventional farm subsidies, and crop insurance, and to simplify regulatory permitting for farm-to-table food marketing and sales.

As much as Brown would like to see more interest in prioritizing soil health, he does not like throwing around the word *sustainable*. "Why would you want to sustain a degraded resource? I don't want that—first we have to regenerate the soil."

In his experience, the information that extension agents, universities, and industry feed to farmers is geared toward promoting chemical-intensive monocultures. He summed it up pretty simply. "Everyone who advises farmers is trying to sell them stuff or make a living off them. People come here to see my equipment! But it's not the equipment that matters. It's the system you use it in that matters." While he may not outyield his neighbors, Brown assures me that his farm is much more profitable. "The guy with the record corn yield lost money doing it—how stupid is that?"

Yet regenerating soil fertility offers benefits that extend well beyond revitalizing small farms. "Have you heard about the drought in California?" he asked. "Or all the nitrates in the Gulf? We've known for half a century that half the nitrates we apply to our fields don't get into our crops." Nitrate pollution, excess runoff, drought susceptibility, soil degradation: these things can all be rectified if we manage our farmlands to fix one problem—soil health. Brown says this isn't rocket science—anyone can understand it. Research funded by big businesses doesn't tend to focus on low-cost preventative practices. But farmers increasingly do.

After all, there's nothing magic about Brown's farm. He just adopted methods based on the full suite of conservation agriculture principles—and took it to another level with livestock.

TURNING WEEDS INTO BACON

The next morning, the sky was overcast as I tagged along with several dozen farmers from Minnesota, Manitoba, and South Africa who had come to visit the farm. Brown does this several times a week each summer, for several thousand visitors a year. He says he has time to do it only because he went no-till—and because Paul cares for their cattle.

The farmers were there to see Brown's soil. We all followed his Green Bay Packers sweatshirt to a 3-acre field, where green beans scrambled up a scaffolding of six-foot-tall corn. He half-joked that he likes to grow beans and corn at the same time because then he doesn't have to bend over to pick the beans. Various other vegetables were growing between the rows of corn; he told the group that this field produces three times as much food per acre than growing a single crop. In fact, they grow enough in this field to harvest their own food, supply their farmers' market demand, and still make substantial donations to a local food bank.

Before he showed the assembled farmers his soil, Brown lined three buckets up in front of them, each with soil from one of the fields he and I had visited the day before. One by one, the farmers shuffled by, thrusting their hands into buckets to feel the soil, break it up, and, of course, comment on it. "Not very crumbly. Not much organic matter." "This one's got no pore spaces." "That's the worst one," opined one farmer after inspecting a clod of tilled organic soil. Everyone agreed that the densest was the high-input one. Nobody was too thrilled with any of them.

During their inspection, Brown had sent a volunteer into his cornfield to get a shovelful of soil. When the volunteer returned, the tone of the commentary shifted as the shovel was passed around. "This is nice stuff," one gentleman standing beneath a cowboy hat said. "It's darker and full of holes—like Swiss cheese," added another. "Nice aggregates," said a third.

Picking up on this last comment, Brown told the group that it takes mycorrhizal fungi to make this kind of soil. Both tillage and too much nitrogen kills mycorrhizae. The former slices them up, and the latter makes plants cut way back on their exudate buffet, which starves the mycorrhizae and other beneficial soil microbes. This was the problem with his neighbor's soil—the problem with conventional agriculture.

For emphasis, Brown held up two chunks of soil, one from his field and one from the tilled organic field next door. The neighbor's dry light-brown soil looked anemic next to Brown's moist blackish soil full of holes, recognizable bits of organic matter, and stable aggregates. The farmers were riveted as Brown spoke. "It's the biology that drives the system," he said. "We all know that. More carbon and aggregates mean more biology and more cycling of nutrients." He'd taken this plot from less than 4 percent organic matter to over 10 percent in just six years, adding about 1 percent a year. He did it without inputs, herbicides, or amendments, just using cover crops, manure, compost, mulch, and his secret ingredient—wood chips to add carbon and boost growth of mycorrhizal fungi.

The big question now, he explained, was how to scale it up, since they didn't have enough wood chips to cover the entire ranch. The native prairie soils had up to 8 percent organic matter, and Paul aims to bring their whole ranch up to 12 percent over his lifetime. How were they going to manage that, the assembled group wondered?

"With help from our cows," Brown replied.

The aboveground chewing, tearing, and trampling by livestock grazing creates wounds that the plant must heal. But the plants don't do it alone. They need soil micronutrients and microbial metabolites— both of which will be delivered only if they pump a steady supply of carbon-rich exudates out of their roots to recruit microbial assistants. This is nature's way of building fertile grassland soils. It's also how cattle can contribute to restoring soil health and promoting nutrient cycling on farmland.

While Brown sees grazing as an effective way to get more carbon into the soil, he doesn't like to say he's "sequestering" carbon. It's simply part of the natural cycle: as plants fix carbon and push it out of their roots to feed microbes, they build up the stock of carbon in the soil. So he uses high-density, low-frequency grazing to restore both his soil and the native prairie.

Brown led the group to an idyllic spot where several hundred chickens scurried around a couple dozen cows grazing on a field of sorghum, squash, and rotting pumpkins. As annoyed cows chased nimble chickens, he explained that this is how stacking enterprises boosts soil fertility—and farm profits.

A pair of trailers labeled "Paul's Poultry" were parked just outside the single-wire electric fencing at the edge of the field. These egg mobiles, as Paul calls them, are made from old cattle trailers. Inside,

Paul had built roosts and little boxes for the chickens to lay their eggs in, and he'd installed grate flooring so that their droppings could fall right to the ground. By day, the chickens wander around and manure the field near their portable coops, then find shelter within them at night. Plus, they trade fly control for the cows for protection from coyotes and foxes, through their proximity to the big bovines. While the farm does lose a few chickens to predators, Paul shrugs that off as part of nature's cycle.

The Browns see the key to making money on a farm as using the cull of one thing as the input for another. They make chicken food out of grain screenings, turning something that would otherwise go to waste into more chickens and eggs. The only time they come out to tend the chickens is to collect eggs and move the egg mobiles around to follow the cows. At peak production in spring and summer, they sell more than $700 worth of eggs a week. They can barely keep up with demand.

Paul has been a driving force behind increasing livestock diversity on the farm, adding sheep and hogs to the mix as well. "Anytime you can turn weeds into bacon, it's a good thing," he said. And their cows don't just graze on native prairie. They also graze on cover crops and harvested cropland.

Cover crops bridge the Browns' no-till cropping and grazing systems. Gabe seeds cowpeas and other cover crops along with their corn, with the corn coming up first and the cover crops then growing in under it. After they harvest the corn, the cattle graze the cover crop and manure the field before no-till planting into the residue.

They'd tried using a crop roller, like Jeff Moyer uses at Rodale, but the cover crop has to be flowering for that method to work properly, and their growing season is too short to get in a cash crop after that. So they use cattle instead, taking advantage of how cows move themselves around—and poop a lot.

Brown will tell you that there's no simple formula for deciding how many species should be included in a cover crop mix. He gets better results with seven or more species and now typically uses ten to twenty, although he has included as many as seventy in a mix. He customizes each mix to his goals for particular fields, mindful of synergies such as combining sunflower and daikon radish to provide deep taproots with buckwheat for phosphorous scavenging. He'll plant what I came to think of as bovine salad mix—winter cover crops that stick up through the snow for their cattle to graze from December into February.

When Brown first got the farm, he seeded some of the degraded soil back to perennial prairie. It didn't work. He had better luck after using cover crops and grazing to start improving the soil. Then he could reseed it to native grasses. After that he began rotational grazing to promote exudate production and, he hoped, further improve soil health.

At first, his farm had three large pastures where cattle grazed year-round. Now it has over one hundred small pastures chock-full of over a hundred species of native grasses that will be grazed for a few days a year, at most. Brown explains the idea crisply. "Cattle have four legs for a reason—so we let them use them." They mix up the timing of grazing so as to be unpredictable in terms of the sequence or time of year. Their cows can graze on crop stubble into February before they shift to bale feeding. Brown won't export baled hay from his land. His cows eat it in the same field it grew in. He will, however, import straw from neighbors when they'll sell him $30 worth of nutrients and only charge him $5 for the straw they come in.

Brown began rethinking his grazing practices when he came across the controversial ideas of Allan Savory, a wildlife biologist turned alternative-grazing guru in Africa. Originally an opponent of grazing, this guru of "holistic management" had made a surprising observation while studying wildlife management in Africa. It wasn't having too many grazing animals on the land that led to soil degradation, it was keeping them too long on any one area. So Savory prescribed intensively grazing any one piece of ground for just a day or two per year.

When I asked Brown his opinion on what drove the division of livestock and cropping in the first place, he was quick to respond. "Extension agents," he said. "The reason feedlots were started was to get rid of excess grain production." And as automation spurred the growth of centralized feeding operations, it saved the farmers who specialized in grain production the extra work involved with livestock. The loss of manure then helped push grain farmers into fertilizer dependence.

Back when Brown and his son first changed up their grazing practices, they didn't see many dung beetles. Three years later, they started to notice more coming around. Now they have sixteen species and Paul has become something of a dung beetle expert. He lit up describing the different kinds: rollers that round up little dung balls, tunnelers that bury dung, and dwellers that simply live in it. Some burrow up to eight feet deep, mixing manure down into the soil. They all eat fly larvae, which helps keep the cows happy.

Changing their grazing practices not only brought native plants back from the brink, but it attracted more native wildlife to their fields. As Brown told me this, a sharp-tailed grouse rose from the field to flee our advance. "It's a great indicator species—if you have those, it's a healthy ecosystem." Brown described how when he first moved onto the farm, he never saw white-tailed deer. But in a recent and particularly hard winter, they found hundreds taking shelter on the east end of the farm. They also now routinely see foxes, weasels, raptors, and "too many songbirds to count." If he wanted to charge people to hunt, he could turn all the new wildlife into another revenue stream. But some he really wants to keep, like the cowbirds, a subspecies of blackbird, that are masters of fly control. He loves that a flock now overwinters with his cows.

Cattle don't get any antibiotics, hormones, or dewormers on Brown Ranch. Eight years ago they stopped giving them vaccinations too. The veterinarian said they were courting disaster, but so far they haven't had many sick cows. Brown explained that cattle get worms when they graze plants low to the ground, so with frequent rotation his don't get them because the plants they eat have had time to grow tall. Their diet of nutrient-dense forage supports their immune system. When calves wean in April, they get moved to graze on cover crops before they either go to market or get reintegrated into the herd. If they've had antibiotics, Brown sells them to a conventional meat producer. He won't sell anything that's been treated under his label.

Brown keeps his herd away from feedlots and other people's cattle. His small herds are like the bands of early human hunter-gatherers whose isolated populations provided natural protection from infectious diseases. Just as big epidemics didn't sweep through human populations until we crowded into cities, cattle tend to stay relatively healthy until crammed into feedlots. We would use far fewer antibiotics and have fewer cases of livestock diseases if we dispersed cattle back onto the land in small groups, with little contact between the herds on different farms. This would also help to keep antibiotics effective into the future.

The first question Paul usually gets from new customers at the farmers' market is "Where are you from?" Then comes "Do you use GMOs?" followed by "What about confinement, antibiotics, or hormones?" Whether the farm is organic is usually well down the list. People are generally happy with Paul's explanation of their low-input, mostly-just-soil-and-sun style of farming. His pitch culminates with "If I was a beef animal, I'd like to live on our ranch."

At first, Brown didn't think there would be much demand for local foods in North Dakota. Now, each week, he loads the market trailer with a lot of eggs and a lot of meat, the equivalent of several cows, a hog, and a lamb. Why are they so successful? He thinks their story—healthy soil equates to healthy plants, healthy animals and thus healthy people—appeals to their primary customers, young families with children who are keenly interested in their kids' well-being and want nutrient-dense food on the dinner table.

On Brown Ranch, grazing starts out on a different paddock each year. An individual paddock will see cattle for a day or two a year. Paddocks that don't get grazed one year are the first to be grazed the next. Forage in the native pasture is remarkably diverse, well over a hundred species. Looking out at the pasture of the moment, I felt like we'd stumbled into cow heaven, with big sky, tall grass, and happy cattle standing around chowing down lush prairie. I have to agree with Paul—this seems like a great place to be a cow.

"So how can we get conventional farmers on board?" I asked him.

"Consumers demanding that we produce good food," he replied. "Talking to consumers is the way to drive change. You're not going to change Washington." I knew he was right. Consumers could drive change faster and more effectively than politicians. Although some changes in agricultural policy, specifically around incentivizing regenerative farming practices, could certainly speed that change.

On our way back from the pasture, a farmer from South Africa asked Brown about the high price of organic food and the problem of global hunger. "It doesn't have to be that way," our host replied. "I can produce food way cheaper than my conventional neighbors. It only costs me $1.44 to produce a bushel of corn. If everybody was doing this, we could produce nutritious food cheaper."

Brown explained that he may not want to get certified as organic, but he's tired of hearing advocates of the organic-can't-possibly-feed-the-world argument compare low-yield, organic, tilled monocultures to conventional monocultures. He thinks that feeding 10 billion people by 2050 will be a challenge only if we keep the conventional production model. His commodity yields are 25 percent higher than the county average, plus he gets ancillaries—cover crops, vegetables, chickens, hogs, cattle, lamb, and honey. He estimates that his system increases per-acre nutrient yields by at least half over those of his neighbors. And he does so at much lower cost.

While he likes that he can charge a premium for his products, he notes that his production costs are the lowest in the area. "On the two key arguments of net production and price, I can blow conventional farmers out of the water. And, in doing it, I have a positive environmental impact. To me this whole argument is childish when you analyze it. No one is sitting down and putting it all together." Letting cattle graze native grass and crop stubble would increase our net food output. That, he says, is how to feed the world.

SOMETHING WORTH CONSERVING

Brown didn't come up with his system overnight. It's the product of decades of trial and error, going back to the 1990s when he and NRCS conservationist Jay Fuhrer began teaching each other a new way to farm.

On the second day of the farm tour, I tagged along to visit Fuhrer at Menoken Farm. Since 2009, the Burleigh County Soil Conservation District has operated this 150-acre farm as a soil-health demonstration. Driving from Bismarck on this bright sunny day, it was clear they've had an impact. Almost all the fields we passed were no-till.

Upon arrival we were shepherded into one of two green-trimmed beige metal buildings where rows of folding chairs faced a long table up front. Fuhrer took center stage and greeted the assembled crowd. Dressed in a checkered shirt, fleece vest, and black jeans, Fuhrer's short-cropped hair peeked out from his broad-brimmed shade hat. Reflective and tactful where Brown is blunt and outspoken, he's in his early sixties and habitually stands with his large hands relaxed on his hips or stuffed into his pockets.

He started off, sermon-like about the need to rebuild fertile fields through getting carbon back into the soil. He reminded us that plants capture carbon through photosynthesis to make their leaves, stems, branches, roots, and exudates. Then he asked, "What does the soil life that eats the exudates do?" Answering his own question, he said, "They build aggregates, pore spaces, and soil organic matter." But the legacy of conventional practices, he went on, has degraded organic matter and disrupted mycorrhizal fungi, resulting in too little glomalin left in the soil. Once the glue holding the soil together is lost, the soil crusts up and water can't sink into the ground. This leads to runoff and erosion.

Fuhrer drew our attention to the front table, upon which were two clods of dry soil. He picked them up and told us that one was from a field that had four years of no-till, the other from a high-disturbance, conventionally tilled monoculture that's never had cover crops. "The high disturbance soil is denser so it should hold up better, right?" he asked. Wrong! He carefully placed the clods into separate coarse-mesh wire baskets suspended in clear plastic containers full of water. The conventionally tilled sample immediately started to disintegrate, slaking soil off into the water. Within minutes there was no soil left in the basket, and the water was a cloudy brown, with dirt blanketing the bottom of the container. But the Swiss-cheese-like no-till soil had remained intact, and the water it was immersed in was still clear. Fuhrer explained that, without glomalin to hold it together, conventionally tilled soil will slake in the first rain. As the farmers crowded around to get a closer look, one blurted out, "Wow, I didn't think it would happen so fast."

After this compelling demonstration, Fuhrer led us outside to a large-scale sprinkler-fed version of the portable erosion simulator that Kofi Boa had showed me in Ghana. A sprinkler head was mounted on a swivel fitting to spray down onto a row of five rectangular deep-dish trays full of soil, each with a different history of cultivation practices. The trays were equipped with plastic containers to collect runoff from the surface of the soil and to collect water dripping down through the soil and out the bottom of the tray. After Fuhrer turned on the rotating sprinkler, the plastic containers told a more compelling story than any PowerPoint presentation ever could.

The tilled, low-diversity, no-livestock soil slaked at the surface, producing lots of brown, sediment-laden runoff. Almost no water filtered into the soil; the container below the tray remained empty. The no-till, cover-cropped soil also produced a lot of runoff, but it was clearer. Again hardly any water infiltrated through the soil. The no-till soil with a high diversity of cover crops grazed by livestock produced a trickle of clear runoff, and a flood of water seeped through to the container below. The soils from both the grazed hayland and native prairie produced hardly any runoff and a stream of yellow, organic-rich water that filtered through the soil into the container below the trays.

The farmer standing next to me was captivated. "Look at that! Amazing. Just look at the difference in colors."

His audience convinced, Fuhrer shifted back into sermon mode. "In our full tillage days, we were always a week away from a drought. You can see why. We weren't infiltrating any water!" Back in the 1980s, they'd tilled and tilled, which led to crusting of the soil surface. The more they tried to bust up the crust the more crust they made. Any rain they got would run off instead of sinking into the ground where plants could drink it up.

For Fuhrer, realizing that tillage wasn't such a great idea after all was just the beginning. When he came to the Burleigh County field office in 1985, he saw bare black earth wherever he went. Everywhere it was full tillage and low crop diversity, mostly small grains, with some sunflower and corn for cattle. Between little organic matter and lots of erosion, he knew the soil was in bad shape. And so he found himself in a pickle—his job was to promote soil conservation, but why conserve an already degraded landscape?

At first, Fuhrer built grass waterways to cut down on erosion in roadside ditches. Then he began to realize that he was simply treating symptoms of a broken water cycle, which was resulting in too much runoff rather than addressing a biological problem rooted in farming practices. He didn't know what he and his fellow farmers should do, but he did know that what they were doing wasn't working. Frustrated and looking for something that didn't yet have a label, Fuhrer went to hear Dwayne Beck speak.

Inspired, he gathered a group of farmers to drive down to have a look at Beck's farm. This trip convinced some to adopt new practices. Gabe Brown was one of them.

Together, an informal team of farmers, government employees, and academics began working to reduce erosion and restore soil fertility. Soon others asked to join the team. USDA scientist Don Reicosky studied how to build up soil carbon. Kristine Nichols looked at the role of soil biology. Entomologist Jonathan Lundgren investigated how diverse cropping systems helped suppress pests and pathogens. With their combined skill set and expertise, they began forging a new view of soil health as the foundation for farm productivity and profitability.

They also began a decade-long push to modify grazing systems. As they experimented with rotating cattle from two, three, or four to a dozen or more pastures, they started to see increased grass and beef production. When droughts came, ranchers could avoid shrinking their herd. This is a major economic advantage for a ranch, since selling livestock in drought years—when everyone else is trying to sell—means the prices will be low. And buying during years of plenty—when everyone wants to stock back up again—means prices will be high. Increased soil health translated into economic resilience. Fuhrer told us how he's heard a number of ranchers say that now more grass is left over when their cows are done grazing than used to grow all year.

As the team began to experiment with cover crops, it bridged the new cropping and grazing systems. Like Brown, Fuhrer thought planting diverse cover crop mixes made sense. "It reminded me of the diversity of the prairie as described in Lewis and Clark's journals." He ordered seed for six cover-crop species—all that his budget allowed.

That spring, Brown and Fuhrer planted monocultures alongside mixes of two, three, four, five, and six species on a 5-acre plot near Brown's place. They were just in time for a bad drought, with less than two inches of rain and temperatures over 100°F. One Sunday, Fuhrer set out for a drive and stopped to take pictures of the cover-crop plots. Much to his amazement, the six-species mix was flourishing and all the monocultured cover crops were dying. He continued to find that the more diversity in a mixture, the better it did. Now, more than a decade later, some farmers in the region have gone to sixteen or more species. Each has a different strength—some, like the legumes, fix nitrogen. Others scavenge nutrients or feed pollinators. The combination increased the nutrient cycling in, and productivity of, the system.

Fuhrer said that the change in practices made his soil more resilient, that crops better endure periods with no rain, and excess rain doesn't pond on the surface but sinks into the soil. Early on, Brown and Fuhrer would talk often about soil moisture. But with this new system, the topic hardly ever comes up.

What have all these changes done for productivity and input use? I asked. Like Brown, Fuhrer has eliminated fungicides, insecticides, and commercial fertilizers at Menoken Farm. He's come to see the wisdom of setting up habitat so the predators have a short commute to get to the pests. He does use some herbicide, due to perennial weeds getting established, but he tries to use a diversity of chemicals so as to avoid developing resistant weeds.

As for the yield question, Fuhrer saw statistically significant increases in yields when they moved to their new system. In the 1980s, they were happy with a 40-bushel-an-acre wheat harvest. Now they get upset if they don't get 60 bushels.

THE GRAZING LINK

In 2007, Brown met Canadian rancher Neil Dennis at a workshop in Brandon, Manitoba. These two short bald guys hit it off, and stayed up half the night talking about livestock grazing on cropland as the way to raise soil health to another level. After visiting Dennis's ranch, Brown was convinced that livestock was the missing link.

And so it was on the morning after the tour of Menoken Farm that Brown, Paul, Greg Friel, and I jammed into a four-seater pickup and headed north on Highway 83 to see Dennis. Freil and I shared the back. A big Hawaiian who manages the livestock at Haleakala Ranch on Maui, he's a man of few words—and those he does use mostly involve cows or grazing. While Paul drove, Brown took a steady stream of calls on his cell phone from people wanting advice or to buy cover-crop seeds.

Driving north over gentle ups and downs, we cruised through a sea of just-harvested wheat, standing corn, and low soybeans. Patches of open prairie and scattered stands of trees flashed past. By the time we got to Saskatchewan, I realized we hadn't seen a hill all day.

After 245 miles, we turned left onto a gravel road between fields of sunflowers in what could reasonably pass for the middle of nowhere.

We arrived at Sunnybrae Farm, pulling up in front of a two-story white house with a front porch sporting moose antlers and a big Canadian flag.

An ex-hockey player, Dennis came out wearing a blue Wawota Flyers shirt, black shorts, and muck boots. Thick black eyebrows capped his bright piercing eyes and matched his drooping handlebar mustache.

Dennis was born and raised on the ranch his great-grandfather homesteaded. Much of his sixty-six years have been spent watching the soil degrade as conventional practices drove him broke. Back in the 1990s, the bank wanted him to get rid of his cattle and go big into conventional cropping, with loans for the requisite new machinery and major chemical use. He thought that sounded like a great way to go broke faster.

When Dennis was a kid, his family had horses, cows, sheep, and crops. Even in the 1970s most folks still had cattle and grain. Then everybody decided that they had to get bigger. So everyone went either all grain or all cattle, and the little towns started to disappear as folks couldn't make ends meet. The middle of nowhere was becoming home to no one.

Dennis's wife, Barbara, hatched a plan to save the farm after she got a flyer in the mail about a course on holistic resource management, which involved rotational grazing methods. Dennis thought it would be easier to go to the workshop than to argue about it. So he went, sat in the front row, and argued with the presenter that it wouldn't work.

Dennis went home and did what the instructor said, determined to prove him wrong. He set up two side-by-side 10-acre paddocks. On one he let his cows graze as usual, and on the other he tried the course-recommended higher stocking density. He was surprised to find that by the next year the plant density on the paddock with the higher stocking density had increased. By the second year, it produced enough grass to graze twice as many cows.

As his pastures kept improving, he kept experimenting, increasing both his stocking density and the rest interval between grazing. By this time he started to realize that maybe the guy was right. His conventional pastures were sprouting weeds and dying back in dry weather, but his intensively grazed fields were lush and no longer required seeding as the native prairie reestablished itself.

Frequent paddock rotation at higher stocking density also altered the grazing behavior of his cows. When they wandered a large paddock, they were more selective about what they ate, seeking out their favorite forage before chowing down. This left the less desirable plants in the field to set seed. In contrast, when cattle grazed in denser numbers, they were not selective about what they ate, grabbing all they could reach before their neighbor did. It's sort of like growing up with lots of siblings, where everything at the dinner table gets eaten—quickly. In the paddocks this helped eliminate weeds and less desirable forage. And when everything got grazed, the nutritious native prairie species bounced back faster.

The key conceptual change was the combination of short-duration, intense grazing followed by a long recovery time. Doing this, the native prairie came back on its own and grew better in dry years than in the wet ones. The trampling action at the higher stock densities created natural mulch on the soil surface, which helped keep soil moisture in place and foster regrowth. The longer Dennis followed this grazing system, the better his grass grew.

And there was another bonus: the longer rest period between grazings broke up the parasite cycle for his cattle. This is because cattle that continually graze on the same pasture give internal parasites a chance to cycle from cows to manure and back as the hosts consume parasite-infested grass. But, with longer rest periods between grazing, there were no cattle present when the parasites needed to complete their life cycle, so they died off. And few to no parasites were present when the cattle did come back.

Before he changed practices, Dennis ran 200 to 300 head of cattle on his 1,200-acre ranch. Now he runs 800 to 1,000, twice as many per acre as those of his more conventional neighbors. In 2006, Dennis produced 152 pounds of beef per acre; his conventional neighbors produced less than 70. Dennis is producing twice what his neighbors do and using no inputs to do it—spending less to make more saved the farm.

It's also saved him a lot of time. He used to need help to move his cattle. Now he moves cows four or more times a day by himself on a glorified golf cart. The Batt-Latch, a programmable solar-powered gate latch, made this possible. Brown and his son use this new device too. It's pretty simple—a solar panel on one side and programmable keypad on the other. It works by spinning a latch on the end to either open or close a gate, thereby automating cattle movement. Dennis's cattle move themselves into new fields when gates open at prescribed times. It used to take him all day to move cattle to fresh paddocks. Now he spends just a couple hours setting out the day's program.

As Dennis's soil improved, the sugar and protein content of his grass increased. The better nutritional profile also saved him money. He reduced the amount of mineral supplements by 90 percent and cut his salt use in half. And so what was a vicious cycle became a virtuous one. As the soil regeneration continues, the higher sugar-content grass pumps out more carbon-rich exudates when grazed, which feeds more soil microbes that help grow more biomass. Over several decades, the carbon content of his soil rose from less than 2 percent to 6 percent, and he says some fields are now up to 10 percent. It seems that Dennis has brought his soil back to its native carbon content, as scientists in the 1890s reported organic matter contents of 5 to 11 percent for Canadian prairie soils. He says his molehills used to be gray—now they're coming up black. And he doesn't get any runoff anymore. The ground absorbs all the rain that falls onto his fields, even in big storms. He told me his fields now infiltrate as much as sixteen inches an hour, whereas his neighbors can absorb less than an inch an hour.

Like Brown, Dennis has seen life come back to the ranch since he walked away from conventional practices. "We've got more birds and wildlife than ever," he said as a squadron of butterflies danced before us. Rotational grazing has opened up grassland beneath the trees in his forest patches, making overwintering habitat for deer. A lot more hawks visit and a fox even follows his cows around to clean up trampled snakes.

We piled onto Dennis's glorified golf cart and headed off to move his cows to a new paddock. Along the way, we crossed the remnants of the old pioneer trail from 1889, faint ruts still etched into the land. Soon we arrived at Dennis's fields, separated by a mix of permanent fences and single-wire electric fences strung between several-foot-high metal stakes with a loop on top. They're high-tech but pretty simple; he can lean out and set a fence line up or take one down while sitting in the golf cart.

A 2-acre paddock with a close-packed mob of 700 cows came into view. They had been grazing it since eight that morning. I'd never seen such a writhing horde of cattle. They'd clipped the grass close to the ground and decorated it with manure piles, leaving a few stems of brushy matter still sticking up. Done with this paddock, the cows won't be back there until next year.

Dennis got out and flipped the latch to open the fence. As if on cue, a temporary leader emerged and the herd flowed behind into the new paddock. Excited by the tall grass, the mob fanned out across the new paddock. One cow, number 1760 according to the red tag in her left ear, came over to check me out across the fence as she grazed. The raft of flies on her back explained the flock of cowbirds taking off like a black cloud from the backs of the herd as we drove up.

Dennis announced that he'll move them again in several hours. In the meantime, this mob went to town on the lush grass. The background noise of crickets and insects faded beneath a low rumble of chomping and grinding. I won't easily forget the sound of seven hundred mouths ripping into foot-high grass.

His eyes fixed on this fascinating scene, Brown told me that the weight gain per cow is better in conventional grazing, but the total gain per acre is better under mob grazing. When I asked him why that is, he explained that with a higher stocking density cows produce a more uniform bite on the plants, more even trampling, better dung and urine distribution, and more uniform hoof impact on the soil.

Mob grazing also packs stubble into the ground. All this helps increase the carbon content of the soil—and thereby its fertility—producing twice the regrowth, which supports more cows.

We drove out to another paddock, where cows were milling around a patch of grassland between stands of aspen trees. They bunched up in close formation as we approached across a field of mixed grasses, yellow, purple, and white flowers, and manure—lots of manure. As Dennis advanced, the cows retreated in step, the calves bringing up the rear of the densely packed herd.

The way the cows moved reminded me of the wilds of Africa, where a lion can trigger a stampede. But in this case Dennis is the lion and his cows know safety comes in numbers. Dennis said that the less the cattle are worked the more they revert to this type of behavior.

High-density grazing with lots of movement is what built the prairie in the first place. Buffalo grazing stressed the plants annually as the great herds moved across the prairie, mowing it down as they went. Under this predictable, regular pressure, plants put a lot of energy into building belowground biomass and a major root system to support rapid regrowth. This translated into prodigous exudate production, supplying go-juice for plant growth *and* health.

Intensive rotational grazing practices have improved soil and forage quality in other regions as well. A 2011 study of commercial ranches in Texas reported significantly higher organic-matter content, nutrient levels, and mycorrhizal fungi abundance in rotationally grazed paddocks than in conventionally grazed ones. And a 2016 analysis found that under regenerative crop and grazing management, cattle could improve soil quality and sequester carbon. The authors suggested combining ruminant grazing with conservation agriculture as a recipe for building up carbon in soils.

Were we to do this in a big way, it wouldn't be the first time that grazing animals helped cool the planet. Over the past 40 million years, grasslands and grazing animals co-evolved and spread to cover 40 percent of Earth's landmass. This global expansion and build up of carbon-rich grassland soils gradually decreased atmospheric carbon dioxide levels. That's how much carbon you can store in a prairie. And this, in turn, helped reduce temperatures, perhaps enough to have helped trigger the geologically recent glacial periods that most know as the Ice Age.

On the drive back to Bismarck, I saw no fences across the vast stretches of cornfields and harvested wheat. Passing through Rugby, the geographical center of North America and the very definition of Middle America, it struck me how this landscape was made for grazing. Herds of buffalo made the rich soil that made this good farmland in the first place. Yet, there are no buffalo there today and all the cattle have been removed from the land.

Brown read my mind. "Look at those cornfields," he said. "Just think how many cows you could overwinter out there." Paul added, "The potential to produce food is unreal."

Brown carried the thought forward. "Most people here feed their cattle on baled hay all winter. Why not leave the cattle to glean through the harvested cornfields or plant some of the hayland with cover crops and let the cattle graze it and return manure to the land? Dave, it just doesn't make any sense to harvest hay and use a tractor to bring it to cattle that have been taken off the land and aren't returning manure to the fields."

I suspected that I didn't need to ask him how he feels about feeding cows corn. But I did anyway. "I've never harvested a cow that had a gizzard. They didn't evolve to eat grain. Grass-fed beef is higher in omega 3s and vitamins. We're burning all those fossil fuels and putting cows into feed lots to create an inferior product." I couldn't argue. It didn't make sense from a lot of angles. Why confine cattle and feed them corn grown with methods that degrade the soil? Why not put cattle back on the land where, as Brown and Dennis show, grazing animals can help build soil and sustain our ability to grow crops?

Contrary to conventional thinking about the detrimental effects of grazing on grasslands, reintegrating animal husbandry onto farms offers a powerful tool for regenerative agriculture—if it's done right! High-intensity grazing that moves mobs of grazing cattle around as a dense herd—the way they naturally bunch up to dissuade predators—can have a positive impact. Manure that we currently treat as waste could, once again, meet as much as a third of global fertilizer needs. If we brought livestock back out onto the land, rural America might be able to support a more dense human population. Re-integrating animal husbandry with crop production on smaller farms can bring life back to farms—and family farms back to life.

Before now, I never would have thought that the key to increasing soil health and food production—both plant and animal—was to introduce more livestock onto the land. The problem that led to gullying of the California Coast Ranges wasn't just grazing, it was leaving no recovery time and concentrating too many cows for too long in the valley bottoms. I realized that we need to move beyond the cows-are-bad mentality to ask instead why we are raising our cows so badly.

Sitting in the Bismarck airport waiting for my flight home, I took a moment to gaze out at the fields surrounding the airfield. What a shame, I thought, that there are no cows.

10

INVISIBLE HERDS

A person with a new idea is a crank until the idea succeeds.

—Mark Twain

After floating another few miles downriver, we pulled our supersize dugout canoe over to the right bank and scrambled ashore. My boots sank into the mud as I cradled the soil auger, a meter-long hollow pipe with an open-center, screw-like drill bit on one end and a handle to turn it with on the other. The jungle was thick, but we were able to clear a path as we hiked out into the forest. When the GPS unit said we'd reached our destination, I pushed the auger into the earth and put my back into drilling.

Once the handle reached the ground, one of my compatriots helped me pull the auger blade back up out of the hole. We popped the clear plastic sampling tube out of the auger and examined the river-deposited sand and silt.

Memories of the day before cast a shadow over our walk back to the boat. One of our sandal-wearing guides had cursed, then stopped short near the end of a transect. The snake that had just bitten his foot scurried off into the brush before anyone got a good look at it. Was it poisonous? we'd asked anxiously. Wait a few minutes, he'd said, and we'd know. A while later, nothing drastic had happened and we decided to move on to the next drilling spot, a dozen relieved, watchful eyes glued to the ground along the trail.

I was on a monthlong University of Washington expedition down the Mississippi-sized Río Beni, floating 500 kilometers from the edge of the Andes toward the Amazon across a vast lowland jungle. Another professor, two students, and I were there to study how the river meanders across its floodplain and determine how often big floods deposit fresh sediments. To do this we needed to collect sample cores to take back to Seattle to date. And that gave us the perfect opportunity to take a good look at the soils beneath the forest floor.

On the whole, we found hardly any organic matter in the soil. Just a thin layer of leaf litter covered the ground surface. In the jungle, dead things don't last long—legions of ants and other organisms quickly break down anything that falls to the ground. Then nutrients get cycled rapidly back up into the aboveground world of living plants.

After several weeks on the river, we pulled into the first village we'd seen in the whole trip in order to restock our fruit supply. Walking through the village, I noted all the organic matter lying about, particularly the discarded remains of fruit and charcoal and ash from cooking fires. Was this why the soil in the village was darker than what we'd pulled up in our core samples from the surrounding jungle?

When the first European explorers came through the Amazon, the native population was large—and intimidating. In some spots tens of thousands of inhabitants lined the river to gawk at the equally curious, and very nervous, explorers. Compelling evidence for long-term occupation by large native populations is recorded in the region's rich black *terra preta* soils.[25] Over thousands of years, the practice of returning organic matter to the soil, in the form of charcoal, ash, bones, and human waste, built up fertile soil in the parts of the Amazon with the most people.

A key ingredient in making these rich soils was biochar, charcoal produced in continuously smoldering fires. In this way, native Amazonians built up nutrient-rich soil around their villages in an environment that otherwise wouldn't have much. Compared to the native soil, terra preta has three times the organic matter, nitrogen, and phosphorus. Today, rice and bean yields on terra preta can be twice those from adjacent soils. Some enterprising Brazilians even dig up terra preta to sell to farmers and urban gardeners. The practices that created such soil were not restricted to South America; recent archaeological excavations in northern Germany revealed deposits of Nordic black earth formed in a similar manner.

And there is a modern version of terra preta production. Use low-oxygen combustion to make biochar and add microbial inoculants to restore fertility to the soil. Proponents see biochar as a way to cultivate microbial life in the soil and help restore degraded farmland in both temperate regions and the tropics.

Art Donnelly, an acquaintance who designs clean-burning, low-cost cookstoves that make biochar, sees the potential for biochar to help farmers around the world build healthier and more carbon-rich soils. It's a key ingredient for fertilizers and amendments that they can make themselves. He'd agreed to introduce me to Costa Rican farmers who use biochar and microbiology to help restore their land.

As afternoon storm clouds gathered overhead, purple hills covered in bougainvillea slipped past along the freeway. The gorgeous vines scrambled up slopes rising above the deep roadside ditches built to handle San Jose's tropical downpours. Houses with rust-red corrugated metal roofs lined our way toward the old capital of Cartago. There we turned off the four-lane Pan-American Highway and wound our way up into the mountains, into the cloud forest where puffy shadows hid the ridgetops. We were entering coffee country.

Donnelly is almost sixty, compact, with a red goatee, and a fair complexion. For thirty years, he did custom metal-art work for Seattle-area restaurants, nightclubs, and high-end residential clients like Bill Gates and Howard Schultz. When business dried up after the crash of 2008, he helped start a nonprofit focused on promoting biochar and clean-burning cookstoves. As he drove, he told me how he became curious about biochar—and how Central American coffee farmers have become some of the biggest agrochemical users in the world.

His interest in cookstoves and biochar began on a visit to the indigenous Kuna people in the San Blas Islands off the coast of Panama. For centuries the isolated islands were a haven for pirates, and when Panama became independent in 1903, the Kuna did not want to be part of the new country. After a 1925 revolt, they negotiated semi-autonomous status. When Donnelly visited the island in 2007, he was shocked at the number of children with respiratory diseases from chronic exposure to smoky cooking fires inside their huts. He also saw that siltation caused by erosion from slash-and-burn farming on steep coastal hillsides was killing the reef and its fishery. With their fisheries and agriculture failing, the Kuna were increasingly forced to trade dried coconut meat used to extract coconut oil for canned food from Columbia. He thought there had to be a better way.

His lightbulb moment came when he was reading about biochar for the first time. Donnelly thought that clean-burning, charcoal-producing cookstoves could help rebuild degraded soils and keep smoke out of the lungs of the kids that he'd seen in Panama. He started tinkering with stove designs and eventually built a prototype. These first simple biochar stoves were a hit with Seattle gardeners interested in making their own biochar. In September 2009 one of those stoves caught the attention of Arturo Segura, a Costa Rican coffee farmer at an organic harvest festival in Seattle.

At the time, Costa Rica's organic farmers were beginning to embrace the Japanese practice of making bokashi, a fermented compost made from organic wastes combined with crushed charcoal.[26] Unfortunately, the charcoal was coming from cutting down primary forest. Segura encouraged Donnelly to visit Costa Rica and show coffee growers how to make biochar from organic wastes produced on their own farms. He also told Donnelly about the impact on respiratory health that exposure to smoky cooking fires was having on indigenous coffee-bean pickers. In January of 2010, Donnelly traveled to Costa Rica to kick off what turned into the five-year Estufa Finca (farm stove) project.

Donnelly's metal-working and welding experience came in handy. He told me that the basic design of a biochar-producing stove is based on controlling the airflow to the fuel chamber to create an oxygen-restricted environment within the stove. The volatiles, driven off the fuel by the top-lit starter fire, burn in a separate secondary-air zone. Once the volatiles have been consumed, the fire goes out, leaving the carbonized remains of the fuel load behind. The stove can use a wide range of dry organic matter to produce an almost smokeless cooking flame—and clean charcoal.

Donnelly began surprising crowds at cooking demonstrations by firing up his stoves for hours at a time—sometimes indoors—with no visible smoke. An independent emissions laboratory tested the stove and determined that it produced hardly any carbon monoxide. His clean-burning stove also produced high-quality charcoal.

A number of factors affect the quality of biochar, the combustion temperature and the quality of the starting organic matter chief among them. Lower temperatures produce more biochar. Higher temperatures produce a biochar with higher pH and higher nutrient availability (cation exchange capacity). But burn too hot and the porosity starts to collapse. The optimum temperature for making biochar is about 500°C (900°F), and Donnelly's stoves hit that temperature and can char pretty much anything organic. He once showed me a whole charred black rose that looked like something out of a Tim Burton movie. But crop residue is the most likely input material for small-scale farmers in the developing world to turn into biochar. With a biochar cookstove they can use their crop wastes for cooking fuel and return the char to their soil.

Biochar can persist in the ground for centuries to millennia. That's why archeologists date ancient sites by radiocarbon-dating charcoal. Yet how can biochar influence soil fertility if charcoal is so inert that it can last for thousands of years? It comes down to its high porosity, which increases the soil's water holding and cation-exchange capacities, and its ability to increase soil pH, which reduces soil acidity to levels that favor beneficial microbial life. It also provides pore space and habitat for microbes that break down and recycle organic matter.

A number of studies demonstrate that biochar boosts both microbial activity and crop yields. In small test plots that mirror such studies, Donnelly saw a 30 percent increase in crop yields on poor soil and no increase from adding biochar to already fertile soils. Field trials in carbon-poor Amazonian soils have shown charcoal additions can more than double crop yields. And a study in French Guiana found that a combination of manioc peel and charcoal provided a bigger boost to both crop yields and earthworm activity than either soil amendment did alone. However, some studies have reported that applying biochar to already fertile soil can actually decrease yields.

At the crest of the mountains, we passed a big wind farm where hundred-foot-tall, triple-blade turbines spun in the Pacific winds spilling over the Continental Divide. As we entered the clouds, diffuse green light and vine-covered trees contrasted with the rust-red soil exposed in the roadcuts. We turned off Highway 2 at Empalme and headed west on a narrow, winding, and very well-maintained road into Los Santos, the heart of Costa Rican coffee country.

It was pouring when we arrived in the small town of Santa María de Dota. The picturesque town filled the bottom of a steep-walled valley, with coffee plantations covering the lower slopes and forest on the steep ridges above. A few landslide scars ran down the valley walls and a pair of enormous red and white metal microwave towers rose high above the valley floor. The modern, warehouse-sized coffee cooperative on the main road sat a couple of blocks from the traditional town square. Promptly at 6 P.M., thunder joined the rain to herald the tropical sunset.

MICROBIAL BREW

The next morning, the sun shone brightly as we set off to visit Felicia Echeverría at her small farm in San Pablo, a few miles west. Along the way, roadcuts exposed gray volcanic rock fading upward into orangeish weathered rock and clayey subsoil with little topsoil and no organic matter on top. Everywhere I looked, red dirt peeked out from bare exposed ground. Even the river was flowing orange-brown after the previous night's rain.

Across the valley, rows of coffee bushes inscribed dotted contours across the hillsides, each bush pruned to the height of a person to keep harvesting simple. Donnelly says the coffee farmers complain that yields are going down, even though they're applying more fertilizer than ever before. He thinks they know they are on an agrochemical treadmill. "Everybody sees the writing on the wall. All the growers know it won't be viable for much longer."

I could see why after we turned right at the church in San Pablo and headed up a narrow paved road that proved a struggle for our car to climb. All the roadcuts told the same story. We stopped at one so that I could photograph it and read for Donnelly the classic tale of hillside agriculture eroding off the topsoil. Clayey blocks of reddish subsoil extended right to the ground surface. There was no topsoil, no organic matter cover. I saw soil leached of available nutrients and enriched in the insoluble iron and aluminum oxides left behind as warm rainwater percolated through the ground. No wonder these farmers need a lot of agrochemicals. Their soil is shot.

A third of the way up the ridge, we left the paved road, bouncing along a dirt track past small farms, patches of forest, and more coffee fields. Donnelly looked for Echeverría's place while trying to avoid getting stuck in the slick red gumbo of wet strips of the road. Still thinking her place must be just ahead, we reached the ridge crest, where a patch of native forest stood preserved. On the outside of a bend, the two-foot-tall roadcut exposed the forest soil: six inches to a foot of dark organic-rich topsoil on top of red subsoil. Finally, here was fertile soil much like what this land had when first cleared for farming—loose, crumbly black-brown silt capped with organic matter. Right below this fragile layer was the hard blocky subsoil now exposed at the surface on farms all around the valley. Less than a century of coffee cultivation had destroyed the soil and the region's native fertility.

Realizing that we must have already passed Echeverría's place, we retreated back down the mountain. We eventually found her waving us down from a tidy fenced compound as we approached from the wrong direction. She greeted Donnelly like the old friend he is and invited us into her small one-room wooden house. One side had a blue-and-yellow-tiled kitchen with a large metal worktable where she makes the salad dressings she sells for a living. Her farm grows all the herbs this venture uses—and most of her food.

Echeverría was in her fifties, thin with a curly halo of black hair. Clad in jeans, muck boots, and a white shirt, she was a bit taller than Donnelly. She had a light step and a warm intelligence, and something about her reminded me of the Frida Kahlo we know from the artist's self-portraits.

From 1999 to 2006, Echeverría led the organic program at the Ministry of Agriculture. She bought her own 4-acre farm after a divorce and moved there in 2010, once she'd saved up enough to build the house. She not only wanted to restore her farm's degraded soil, she wanted to show farmers that a single woman in her fifties could do it—and convince them that they could too.

The first thing she wanted to show us was her dry composting toilet, which uses a single cup of water per flush. She'd seen too many rivers and lakes in Central America that smell like human waste and thinks that people need to do their business differently. With no tank, her toilet uses just enough water to lubricate material going down. A squirt from a short hose hanging on the wall can move any stubborn stuff along. Rainwater collected off the roof supplies the toilet, which drains to a tank full of wood chips where solid material collects. She turns it once a week, periodically adds more wood chips, and empties it just once a year.

She then mixes it all with kitchen scraps and lets it compost for a couple of months before applying it back to her banana trees or flowers. "I don't think of this as a fertilization system. It's a way to reuse your waste." Her kitchen water and toilet water pass through an underground filter of rocks and soil before flowing out onto the slope to water her flower garden—but not her herbs, she's quick to point out. She showed us her latest batch of finished compost. There was no odor, and it looked like rich soil.

Walking through her herb garden, I couldn't help but notice hummingbirds big and small darting about. The terraced hillside was well stocked with dill, lemongrass, rosemary, lavender, and a host of herbs I didn't recognize.

The second thing Echeverría wanted to show me was back in the octagonal, open-air, cement-floored structure in front of her house. There in large blue and white containers, she makes a bokashi leachate using MM (EM-may EM-may), shorthand for *microorganismos de montaña* (mountain microbes). This low-tech microbial inoculant employs a sourdough-like starter to brew up a biofertilizer rich in mycorrhyzal fungi. To get it going, Echeverría collects a bag full of organic matter (humus) from the forest floor, making sure that it contains white fungal hyphae. Next, she removes all the big leaves and mixes it up with rice bran and molasses. She then covers it with a plastic bag in 60-liter, screw-top containers and lets it ferment for a month and a half.

She carefully unscrewed the top of a container and I peered inside. A faint, sweetly pungent odor drifted up to meet my nose. The resulting product boosts soil fertility and plant growth and health—if made properly.

This fine, granular stuff also serves as a starter for aerobic microbial tea. To make this, Echeverría submerges a bag of the solid material in water mixed with molasses. After steeping for two weeks, this turns into a fungus-rich brew of organisms cultivated from forest soil. She applies the fermented solids to seedlings and her planting beds, and sprays the MM-rich solution on her corn, beans, potatoes, yucca, taro, garlic, oats, and other crops to inoculate them with beneficial microbes.

"We use this all over Latin America because it is cheap and it works," she told us. But she quickly added that you have to be careful to recognize a bad batch of MM, in which the wrong fungus has grown and, if used, could harm plants. To emphasize her point, she unscrewed a container holding a batch that had gone bad. I was immediately accosted by a sulfurous stench rising from the anaerobic muck inside. If that's the test, it seems pretty easy to tell when it's gone off. Still, it's hard to standardize a biochemical process that starts with a shovelful of forest soil. And because it's not reliably repeatable, it invites skepticism among scientists despite farmers' experience that it works—most of the time, anyway.

Nonetheless, Echeverría showed me a Ministry of Agriculture booklet that spells out recipes for using this approach to make manure without a cow. Not all farmers have access to livestock, but they all have access to soil. In her experience, the recipe for MM is easy, cheap, and enhances soil fertility. Just fill a large container, such as a 55-gallon drum, with six parts forest soil (humus) to four parts rice bran and a gallon of molasses, mix it all up, seal it, and let it ferment anaerobically. When brewed correctly, this makes a good fertilizer that can also serve as a supplemental feed to introduce beneficial microbiota into chickens and cows.

Echeverría also makes fermented compost that includes about 10 percent biochar. This practice of making bokashi to cultivate beneficial microbial life was introduced to Costa Rica in the 1990s by a disciple of Dr. Teruo Higa, who modified a traditional Japanese practice he'd learned from his grandmother. It caught on enough with Costa Rican farmers that the Ministry of Agriculture published a recipe for making it. It looks something like this: mix chicken manure, sugar-cane waste, liquid or solid MM, charcoal (or ash), molasses, rock dust, and old bokashi. The agency particularly recommends it for spurring seedling growth.

I was also intrigued by Echeverría's recipe for an all-purpose pesticide, homemade from ingredients that, except for the MM, would double as tasty additions to a good lunch: garlic, hot chili, onions, ginger, molasses, vinegar, alcohol, and water. After fermenting for fifteen days, the mixture can be diluted and sprayed to control insects, nematodes, fungus, and other pests. It sounds like it would also make a fine hot sauce.

Echeverría describes herself as a biointensive farmer, someone who tries to grow everything she needs to feed herself and her soil. "Biointensive is like biodynamic without the weird part. It's much more normal—farmers here would not follow the voodoo stuff." She ignores the spiritual and mystical aspects of biodynamic farming and its talk of cosmic forces in the soil and simply plants according to what she likes to eat and what makes good compost.

We marched a few hundred meters up the dirt road above her house to her fields, past the neighbor's highly fertilized conventional coffee plants rising out of crusty red dirt. Along the way she explained the basis for her biointensive system: keep the ground covered and use mulch, mixed crops, and inoculation with MM solution. She doesn't use chemical fertilizers or pesticides. To prepare a new planting bed, she digs down two feet to break up her blocky soil. Then after planting, she doesn't disturb the ground. This sounds a lot like the principles of conservation agriculture used by Beck, Boa, and Brown—with the addition of microbial inoculation to kick-start an invisible herd of microscopic livestock.

Echeverría said that her original soil was so hard she had to use a pickaxe to break it up and so clayey she used it to make pottery. Determined to show me what she was up against, she grabbed a pitchfork, walked over to an unplanted edge of her field, jammed the business end into the ground, and jumped up on its flat bar. It went nowhere. She jumped up and down on it some more. It still went nowhere. No wonder the prior owner abandoned the farm.

When she bought the place, it was covered in wild blackberries, a few trees rising through the brambles. Now she has twenty-two planting beds nestled between fruit trees. She also planted nitrogen-fixing trees and others that produce nitrogen-rich leaves ideal for composting.

Initially, she had a hard time making a go of it. In her first year, she burned up all her savings, ran up $2,000 in credit card debt, and still wasn't producing enough food to feed herself, let alone harvest marketable crops. When she got her soil test back, she found her unproductive soil was particularly low in calcium and boron. A soil-scientist friend interested in biointensive farming recommended using rock dust, micronutrient solutions, pig manure, and extra compost.

Within a few months, she saw major improvement in her soil. Her planting beds produced seven-foot-tall corn, a far cry from her first sickly crop of stunted purple ears. It was all a matter of applying the right combination of rock minerals, organic matter, and, she thinks, microbes to access the nutrients in minerals and organic matter. And getting something to grow in the first place helped spark the process. When she pulled out her first stunted carrots, a halo of surrounding soil was black, a fertile ring slowly growing out into the red soil.

We bent down and easily dug our hands into the now loose, brownish soil of her planting beds. She's still amazed by the changes in her soil and the healthy crops of oregano, rosemary, and mustard seeds that support her salad-dressing business.

Pleased that she could bring bad soil back into production so fast, she began spreading compost and applying MM to her whole farm. In addition to her corn, tomatoes, and beans, she's growing guava, blackberries, peaches, avocados, lemons, and oranges. When I noted that she's basically planted a supermarket, she responded, "Yes, that's the whole idea."

She thinks that her approach can bring a farm back to producing good crops within two or three years. She wants to show other farmers in the region how to make a living out of returning life to the land, and how quickly land can be brought back to productivity—and prosperity. As the rain started back up later that afternoon, all three of us piled into the car and drove down the mountain to meet a coffee grower who was trying to rebuild his soil too.

On the way, Echeverría told us that coffee yields have dropped over the last fifteen years. It used to be controversial crazy talk to say harvests were down because agrochemical farming degraded the soil. But now, she says, it's common knowledge among farmers.

We were a bit early to meet her friend, so Echeverría took us to see Gabriel Umaña, the local agent for the Ministry of Agriculture. We ducked out of the rain into his single-story blue office. Umaña looked like an engineer in his button-down shirt, thick dark-rim glasses, and gray slacks. He was talking with a pair of coffee farmers who were seeking advice about how to test their soil. Even my limited Spanish was enough to catch that one of the farmers uses less than half the chemical fertilizer he used to. The other said that he'd cut his use by almost three-quarters. They both wanted advice on how to further improve their soil and cut chemical inputs (and costs).

After Echeverría conducted a round of introductions, Umaña told us that the pair is not unusual. A lot of small farms have started making big changes—planting more shade trees, composting, and trying to use less pesticide and herbicide. Some were applying more organic matter to their land, and even the town's coffee cooperative was using their coffee pulp to make compost.[27] Interest in such changes had started a decade ago, when the Ministry of Health cracked down on dumping coffee pulp into rivers. Now the cooperative sells composted pulp back to members at good (cheap) prices to apply on their fields.

He's seen farmers starting to warm to the challenge of increasing soil organic matter and cultivating soil life. Increasingly, they see how soils with less organic matter and less microbial activity produce bad-quality coffee and plants more vulnerable to disease. Healthy, fertile soils produce better coffee with less expense for chemicals and lower risk of disease.

Umaña encourages farmers to think of roots as the heart of their crops. And the roots, he tells them, need the good microbes to deliver minerals and cycle organic matter. He tells them that farmers who apply too much herbicide and fertilizer kill off the beneficial microbes and create new problems for themselves. Although he finds that most farmers know little about microbial life, he thinks it is when a farmer begins to understand the relationship between minerals, compost, and microbes that they can really start to manage and improve soil health. So he's trying to teach them.

Umaña recommends a combination of ground-up minerals, composted organic matter, and microbes. It is a lot cheaper than synthetic chemical inputs. Big farms can afford to buy agrochemicals in bulk and are not as concerned with coffee quality. They can always sell what they grow because the cooperative needs to buy from large farmers to get volume.

Why is it so hard to change the big farms? First of all, he said, farmers can't get bank loans to do large-scale MM or bokashi. Another barrier is that there is little institutional interest in promoting methods that have no commercial backing, and it's hard to convince skeptical farmers to adopt a product no company stands behind. And modern farmers tend to see something they can make themselves as old-fashioned and unscientific. Umaña admitted that he is unusual within the Ministry of Agriculture, which mostly steers farmers toward conventional chemical methods.

It takes some time for farmers to start seeing results with his methods. In his experience, coffee farmers begin to realize comparable yields several years after converting to a biointensive approach. Yet farmers immediately notice the significantly lower input costs. He says many notice improved coffee quality within a few years and report that MM synchronizes maturity among coffee bushes. This helps farmers with a major logistical problem, as coffee beans must be sold when ripe. Umaña told us that the hardest part is getting farmers to try his approach and stick with it through the first couple of years. Those who do don't go back.

Next on the itinerary was a meeting with Echeverría's friend Javier Meza. His coffee farm was partway up the valley wall, with a grand view overlooking San Pablo. With his short black hair tucked beneath a baseball hat, he greeted us warmly and told how he's worked this land for more than thirty years. A few years ago he realized that his soil was "no longer any good." After decades of conventional coffee growing, soil tests showed it was depleted of nutrients. Motivated by a mix of economics and a desire for change, he enlisted Umaña to help him improve his soil and restore his farm.

We stood gazing out across the valley. Then I looked down at his soil. The ground lay covered with charred and composted coffee pulp. He plants his new coffee starts through this black ground cover. In addition to putting pulp back onto the soil, he now leaves weeds and grass cuttings to rot when he cuts what grows between his coffee bushes. He's also planted banana trees to provide shade for the coffee.

Microbes are the foundation of his approach. He inoculates the soil around his plants with *Trichoderma*, a common genus of fungus that forms symbiotic relationships with plants and is known to suppress its pathogenic brethren. He also applies *Baccilus subtilis*, a bacterium common in both soil, and interestingly, the human gut. In the soil it enhances plant growth, solubilizes sulfur and phosphorus in the soil, and provides protection from root pathogens. In addition, his plants receive liquid MM and supplemental micronutrients (like boron and zinc) on the foliage and the soil. He's seen big increases in growth within months after he started making liquid MM and applying fermented biofertilizers. He proudly showed us his biofertilizer shed, stuffed full with 200-liter solid MM barrels and 30-liter liquid MM containers.

He's been using these various soil amendments for several years, applying them along with mineral fertilizers, and already he has seen his soil get darker and looser. It helps that biofertilizers are much cheaper to make than conventional ones are to buy. Meza now uses about a quarter of the fertilizer he formerly used, and he is no longer using herbicides. Although he has not gone organic, he has greatly reduced his chemical inputs—without reducing his harvests. He likes the effect on his farm's bottom line.

Soil microbes, like *Trichoderma* species and the caterpillar-killing bacterium *Bacillus thuringiensis,* have been used as inoculants for decades.[28] While microbial inoculants have been adopted in a wide range of agricultural settings, it has proven challenging to consistently reproduce their beneficial effects. Still, studies from around the world show that inoculation with cyanobacteria as a biofertilizer can increase crop yields, often by more than 10 percent. Cyanobacteria influence the availability of phosphorus to plants and can act as bio-control agents against plant pathogens. Likewise, inoculating arable soils with plant growth promoting rhizobacteria has been shown to significantly increase crop yields and help protect crops against disease. On degraded tea plantations in Sri Lanka it took less than a year for applications of microbial communities to improve yields and slash fertilizer use by half.

Meza thinks his coffee plants are healthier since he stopped using herbicides. This is especially important to him now, as many coffee farmers have been starting to have serious problems with "roja," a fungal blight that can wipe out an entire crop. Meza thinks the chemicals and fungicide that have so far controlled the blight on conventional farms have done harm to his neighbors' coffee plants. He believes that healthy soil offers the best frontline defense to this growing problem in Central America. His plants didn't have any problem with the blight. He attributes this to progress in restoring his soil.

The turning point for him came eight years ago, when he quit the local coffee cooperative and started milling his own coffee. Building his own micromill allowed him to compost the coffee chaff the process produces and treat it with MM before returning it to his fields.[29] His coffee processor stands across from his house. It's a small building merged with machinery painted blue and gold, the national colors of Costa Rica.

Inside the building, his automated drier uses the coffee chaff in a gassifier-style furnace to heat his beans to between 50°C and 55°C. Before he built his own mill, he used a lot of wood to run his coffee driers. Now he uses the coffee chaff for fuel. He then mixes the charcoal that this process produces into compost to return back to the soil. His next step will be to bring in worms to help compost the chaff before returning it to the fields. Then he'll start using the worm compost to make compost tea.

Both Meza and Echeverría are working to improve the health of their soil through returning organic matter to it, keeping the ground covered between plants, and introducing and cultivating microbial life. They save money through dramatically reduced fertilizer use, and eliminating herbicides and insecticides.

That night, as we all talked over a pleasant dinner back at Echeverría's farm, I realized that the Costa Rican farmers I'd met had a lot more in common with their North American counterparts I'd visited in North and South Dakota than they probably suspected.

A FOOD FOREST

The next day, Donnelly and I got up at dawn to head off to the Caribbean side of the country, where we would meet up with a cacao grower who, Art said, made exceptional chocolate on his farm. After crossing the forested mountains, I started to notice the color of the rivers and streams we crossed. Two big rivers, Río Madre de Dios and Río Honda, flowed clear, carrying little sediment from their source in a mountainous forest reserve with portentous thunderheads gathering overhead. Small tributaries coming in from the forest also flowed clear until joining Río Pacuare, which flowed orange-brown through a lowland agricultural region. When we reached the industrial banana plantations along the coast south of Puerto Limón (the home of Calypso music), the rivers turned a dirty green-brown. Seeing this, Donnelly commented on the state of local fishing and how it had collapsed after chemical runoff from the banana plantations killed off the reef. I suspected that a high load of silt coming off farms didn't help matters any.

After a full day's drive, we arrived at our lodgings in Puerto Viejo, an old restaurant converted into a guesthouse along the main road. The town sat on a wave-cut terrace, a tectonically raised former seabed stretching from the modern beach to the first ridge inland, the old sea cliff. Puerto Viejo is a sleepy town, where handsome dreadlocked guys on horseback flirt with European backpackers in peasant skirts. The crowded streets were a mix of tourists, expats, Caribbean blacks, indigenous people, and locals of Spanish descent.

The air was stiflingly hot, humid enough to drink. After we settled in, I went down to the beach to watch the sunset and the surf breaking on the shore. Gazing along the shore, I could see a solid line of trees and waves stretching off to verdant mountains in the distance. It seemed timeless, a living postcard from the past.

The next morning we headed off to visit Peter Kring, an American expat who moved here in the 1980s, excited about permaculture and the prospect of designing agricultural systems by emulating natural ecosystems. We arrived at his farm at the base of the old sea cliff, walked across a little bridge, and entered a food jungle.

Kring, clad in shorts, rubber boots, and a loose shirt, greeted us at the small nursery that serves as the front door to his farm and which supplies garden and houseplants to the expat community. After sailing through the South Pacific, he'd landed in Costa Rica, seeking a good place to live and grow things. That was a while ago, judging from his salt and pepper hair, short gray beard, and obliviousness to mosquitoes. He immediately started talking to Donnelly about making biochar.

Kring makes it in 55-gallon barrels, producing 20 pounds per burn that he breaks into small chunks before dousing it with a commercially prepared microbe solution rich in bacterial species of *Lactobacillus* and *Actinomycetes*. He soaks the charcoal in a bucket of water, molasses, and humic acid and then spreads the inoculated charcoal onto his fields. He uses this instead of uncontrolled MM because the controlled mix he buys is inexpensive. "Why screw around? I get a consistent result and it saves time."

Kring landed in Costa Rica in 1986 and the following year purchased an abandoned cacao farm. Since then, he's returned it to productivity, with a serious twist. Now he grows more than 150 varieties of fruits and spices on the farm instead of one. He sells a couple dozen crops but the big cash crop is still cacao—and the chocolate he makes from it. All in all, the farm sells an average of about $800 a week of fruit, chocolate, and plants. He uses no fertilizer, no imported compost, and no herbicide, just biomass (mostly decaying leaves) and biochar. He says that, instead of fertilizing, he tries to promote microorganisms and soil health through keeping the ground covered and growing a diverse array of crops. This sounded like the essence of conservation agriculture to me.

Kring has been inoculating his crops and soil with microorganisms for a dozen years and making biochar for three. He describes his holy trinity of soil care as (1) biocatalysts (microorganisms), (2) biomass (food), and (3) biochar (habitat). He adds biochar to the soil when he plants and then comes back later and adds more. But making full-on terra preta overnight would be overwhelming—he figures it would take about 10 pounds of charcoal per square yard, which pencils out at hundreds of tons for his 45-acre farm. So Kring relies on biomass produced on site for organic matter and uses biochar as a delivery vehicle for microorganisms to enhance soil fertility. He also directly applies the microorganism solution with a backpack sprayer.

Cacao trees produce a lot of biomass, so the ground is never bare beneath them. We walked through a cacao grove, stepping on a thick layer of decaying leaves on the ground. Kring stopped, bent down, and brushed away the leaves to reveal a decaying layer perforated with white strands of fungal mycorrhyzae. A flurry of tiny creatures scurried for cover. He said that conventional cacao farmers will sweep away the fallen leaves. But he doesn't. "I hate bare ground." He sounded just like Kofi Boa.

Following Kring around his farm was like taking a nature walk through a tropical forest. He stopped at nearly every bush and tree to describe the food it produces—here, virtually everything does. Yet, there was room to walk between the trees in this closed canopy food forest. Stepping over fallen logs and sticks in varying stages of decay, Kring said that it takes less than six months for leaves on the soil surface to rot away.

The part of the farm on the marine terrace, below the ancient sea cliff, was divided into paddocks, demarcated by broken-off surfboard tips planted like tombstones. Each paddock was about the size of a football field, with one or two dozen types of crop trees. Every tree produces some kind of fruit—plums, avocados, cola nuts, black pepper, nutmeg, jackfruit, spiny palm fruit, dragon fruit, or durians. Even the dead, downed wood produces food—Kring harvests three kinds of mushrooms from it. Because most fruits are in season for just a month or two, his diversified crop portfolio is essential for ensuring year-round income.

Yet the lack of expense for chemical inputs means that Kring doesn't need to grow as much cacao as conventional farmers do to turn a profit. He can make money on less than a fifth of the harvest his neighbors strive for. So he can do really well financially even if he harvests just half as much as they do. That he can do better with less work is fine by him. It allows him to focus on growing the best quality he can, seeking out old varieties to plant and test.

Kring isn't the only one that benefits. The wildlife seems to do well too. At one point while we were walking through one of his stands, I noticed tiny red frogs hopping through the moist leaves blanketing the ground surface. In the next paddock, we stopped to look down at a stream at the edge of a patch of Krings's food forest and saw a river otter scoot under a log. A big green lizard had run across it just seconds before.

The main stands of cacao and vanilla were up the hill, above the ancient sea cliff. So we started climbing a primitive wood-step staircase where the slope was too steep for harvesting cacao. The trail wound up through a patch of three-foot-diameter trees, which Kring said were less than a century old. Trees grow fast here, and he relies on forest debris to feed the soil. The tallest trees rising straight above the canopy were lumber trees—all the buildings on the farm were built from them.

The diversity on Kring's farm is not just in his variety of crops, but in the different varieties of cacao. He grows eight types that he's grafted and four others that he's collected. The cacao forest floor lay covered with two-to-three-inch-long cacao leaves in varying states of decay, the soil so spongy that you sink into the organic layer as you walk.

I stopped to dig into the decaying leaves. There were loads of fungal hyphae, white stringers and clumps entwined with decaying organics and mineral soil. The topsoil was fluffy, but six inches down I hit hard clayey red subsoil. It was sobering to think that the fertility of the entire forest relies on such a thin fragile zone.

As I got back up from inspecting the ground, I saw a strikingly blue, half-foot-long damsel fly, its dragonfly-like X wings suggestive of a cross between an insect and a *Star Wars* rebel fighter. On my way down the hill to catch up with Kring, I stopped to inspect a colorful pair of inch-long poison dart frogs hanging out at the base of a cacao tree. Three howler monkeys scolded me from the edge of the cacao stand, as I carefully navigated around round holes made by large ground crabs.

A little farther on, I could see the Caribbean in the distance, through the cacao trees at the top of the ancient sea cliff. There, perched at the edge, Kring's treehouse rose on stilts, strategically placed at a natural window through the forest canopy looking out to the sea. Standing on his deck to take in the million-dollar view while enjoying some of his excellent homemade mango brandy, I realized that it would have looked similar a couple centuries ago. How many highly productive farms can you say that about? He's made a food jungle into a successful commercial farm.

UNDERGROUND LIVESTOCK

The common thread linking the practices that helped restore Echeverría's farm, Meza's coffee plantation, and Kring's cacao forest boil down to the principles of minimal soil disturbance and adopting practices that build soil organic matter and cultivate microbial life. The fertility of tropical soils is rooted in a high rate of microbially mediated biomass turnover—while tropical regions have the highest rates of biomass production, they also have the highest rates of decay. So it's hard to build up organic matter. This means that farming in nature's image benefits from cultivating underground herds of beneficial microbial life through providing adequate food (organic matter) and housing (soil structure and biochar).

Bacterial communities in biochar-rich soils differ from and are more diverse than those in char-free soil with the same mineral composition. Biochar shapes beneficial microbial communities and increases microbial biomass by ameliorating the pH of soils acidified by over-reliance on nitrogen fertilizers. It also enhances root mass and growth of mycorrhizal fungi, and thereby increases plant uptake of phosphorus and micronutrients like manganese. And biochar's porosity means that adding even a little to clayey soil helps improve soil texture, water retention, and nutrient-holding capacity. Biochar has also been shown to induce plant resistance to pests, although just how remains unclear. Yet crop-yield responses to biochar can be complex because the source material and preparation methods for biochar influence its effect on pH, crop yields, and soil life. Due to its slow pace of decay, biochar offers long-term carbon storage as well as positive effects on soil fertility that are particularly beneficial in the tropics.

Of greatest significance, however, may be biochar's role as microbial habitat. Soil organisms colonize biochar when it gets buried, much like the way marine organisms find their way to a coral reef. And due to their influence on soil ecosystems and fertility, microbes have been referred to as the chief ecological engineers needed to restore degraded agricultural soils. Even in the microscopic world, free housing is a powerful recruiting tool.

SILVER LINING

Biochar also offers a powerful way to tinker with the carbon cycle and remove carbon dioxide from the atmosphere. As plants grow, they take carbon dioxide into their leaves and use the power of the sun to convert it into living tissue. Turning organic matter into biochar helps store a portion of that captured carbon beneath our feet for a long time, given that the half-life of charcoal in the soil can exceed a thousand years. Estimates of the global carbon offset potential from making biochar range from about 500 million to more than 6 billion tons of carbon a year, or from less than 10 percent up to two-thirds of current fossil fuel emissions. Even the low end of these estimates could help buy time for the world to transition to new energy sources and more climate-friendly farms, cities, and lifestyles.

Biochar and organic matter can also restore soil fertility fairly quickly—in years, not centuries. There is no better ready-to-go, cheap technology to address the multiple problems of eroding, lifeless, low-fertility soils and an atmosphere overloaded with carbon dioxide. Burying charcoal in the soil is an elegantly simple and economically feasible way to help reduce atmospheric CO_2, one that is generally overlooked in favor of technologically complicated—and expensive—ideas for carbon sequestration, like pumping CO_2 back down old oil wells.

There is also significant potential to link energy production to biochar. Here is how it might work. The raw material to produce biochar ranges from food waste and landscape clippings in cities, to crop and noncrop plant materials on farms, to slash from timber operations. Meza's coffee drier is just one example of how to use this material to produce both energy and char to return to the soil. Another example is how a commercial power plant in Denmark uses straw to make biochar. When scaled up to full production, it will produce 10,000 tons of biochar a year, which can be returned to agricultural land to enhance soil fertility. About half of the carbon originally captured in the straw through photosynthesis is retained in the biochar. So the more energy it produces, the more carbon is scrubbed from the sky. Currently, this is the only energy-producing system with a *negative* carbon footprint.

However, there are trade-offs to consider. As we've seen, retaining organic matter from cover crops is essential for mulch and no-till practices. Continual removal of all crop residues to produce biofuels or biochar would lead back to soil degradation. Likewise, it makes no sense to cut down primary forest to produce biochar. But real opportunities do exist for repurposing rural and urban organic wastes to produce energy, sequester carbon, and build soil fertility. The greatest potential for biochar to increase crop yields is on highly weathered, nutrient-depleted soils in tropical regions.

Terra preta, like what is found around Amazonian villages, offers a model for how to build fertile soils wherever there is the potential to combine organic matter and biochar. It makes far more sense to turn organic waste from households, villages, and cities into biochar than to dump it in landfills. Adapting the principles that favored formation of terra preta to modern farming practices is another way to help make soil building a consequence instead of a casualty of agricultural production.

Still, even well-housed microbes need to be fed. And while biochar can help restore fertility to soils, it won't offset the need to grow cover crops of green manure on farms. And this led me back to the heart of my own country. For Ohio, it turns out, is one of the best places to see the transformative potential of cover crops for American agriculture and how carbon farming could have a big impact on both agricultural production and climate change.

11

FARMING CARBON

I bequeath myself to the dirt to grow from the grass I love,
If you want me again look for me under your boot-soles.
—Walt Whitman

Seven years after I first met Rattan Lal in Washington D.C., I found myself sitting in his office at Ohio State University. I'd come to see his long-running experiment on how much carbon no-till farming could put back in the ground. At least that's what I aimed to find out.

We jumped right in, talking about long-term food security and how legacies of past soil degradation have set the stage for regional strife and humanitarian disasters. Looking forward, we'll need all billion and a half hectares of Earth's cropland to be as fertile as possible if we are to reliably feed several billion more of us later this century.

Still spry at seventy, Lal became animated when talking about the power of mulch. On his desktop monitor he pulled up photos of a pair of Ohio cornfields during the 2012 drought—one field mulched, the other not. To provide scale, a six-foot-four student stood amid the corn in both photos. The plants in the unmulched field came up to his belt; in the mulched field, much greener corn reached up to his eyes. Lal's smile said it all: Behold the power of mulch. Examples like this have convinced Lal that conservation agriculture can raise African crop yields and sustain intensive agriculture around the world.

As the conversation shifted to the results of Lal's early plot experiments in Africa, it became clear why he is known for his depth and breadth of knowledge about conservation agriculture. Not only did he do some of the foundational work in the field, but beside me stood a large, six-level bookcase stuffed with books on the subject. I didn't notice at first, but when he got up and pulled a tome off the shelf to show me a graph he was describing, I realized they all shared a common author—Rattan Lal.

He'd clearly been keeping busy since he'd returned to Ohio State in 1987. But his research focus shifted after he began finding it difficult to obtain support for basic research on soil erosion and practices that did not involve marketable products. Since then, he's kept his experimental plots going in a decades-long study of the potential for farming practices to build up soil carbon and thereby reduce the amount of carbon dioxide in the atmosphere. Along the way, he's found that the recipe for tempering the problem of erosion and enhancing drought resilience—don't disturb the soil, mulch crops, and grow variety—also offers a potent combination for tackling climate change.

And whether researching solutions for agriculture or climate change, he's disappointed when influential scientists, respected friends and colleagues, endorse gambling on yet-to-be developed technologies instead of encouraging wider adoption of practices already shown to work.

Part of the problem is that policymakers and scientists alike gravitate to silver-bullet fixes and high-tech solutions, like pumping CO_2 deep belowground in the case of climate change or splicing new genes into plants in the case of agriculture. To illustrate this point, he launched into a story about how the Department of Energy shot down funding for soil sequestration work in a big project on which Lal was one of the principal investigators. The program started out with around $100 million for studying carbon sequestration in soil and studying deep borings drilled into bedrock. But the agency folks wanted a big project they could point to—50,000 tons of carbon in one hole makes a bigger impression than half a ton per hectare spread across 100,000 hectares of farmland. So Lal's hand in the project got dealt out of the deck, and all the funding ended up going to investigate pumping CO_2 down deep wells at coal-fired power plants. Meanwhile, Lal's low-tech approach to distributing on-farm carbon storage languished.

As much as Lal likes to talk about the potential to sequester carbon in soil, he'd rather show me. So, I fell in line when he rose and donned a wide-brimmed floppy hat emblazoned with the logo of the International Union of Soil Sciences. It was a fitting piece of headwear, given that he was their newly elected president. I followed him out to the parking lot, trailed by two of his staff, Jose Guzman, a postdoc from Kansas, and Basant Rimal, former undersecretary of forestry in Nepal.

We got into a silver Ford Fusion and left the university setting for what looked like a farm on the far edge of campus. When the car almost got stuck on the bumpy dirt road between rows of various crops, Lal joked that maybe we should have taken the truck. Soon we pulled up to his no-till research plots. They looked like any number of cornfields I'd seen over the past few months—except that in this one some rows were a lot taller than others.

The OSU no-till plots were part of an ongoing experiment, run with three distinct treatments since 1990. From the start, each row consistently received the same amount of nitrogen, though the source differed. Some rows received chemical fertilizers, others compost, and yet others cow manure. So differences in plant growth and health were not due to the amount of nitrogen added, but to *how* it was added.

Even a casual look revealed that crops were not equal after twenty-five years of cultivation under these treatments. The corn in the manured rows was a third taller and darker green than the conventionally fertilized rows. The composted plots were in between in both color and size.

The differences belowground were even more pronounced. Guzman and Rimal used a soil auger to drill up a core from the chemical fertilizer row and the manured row. The soil in the conventional control plot was light brown, with a dense, clayey feel. The loose-feeling soil in the manured row had a cap of plant litter and was blackish with organic matter. Looking at these cores side-by-side, I didn't need fancy statistical tests to tell that adding compost or manure improves soil structure and increases soil carbon.

Yet never is carbon laid out as an integral part of fertilization plans. Usually farmers and researchers focus on nitrogen, phosphorus, potassium, and maybe calcium, sulphur, or zinc. That's because plants don't directly take up soil carbon. But it feeds their microbiome, the menagerie of microbial alchemists in the rhizosphere that, by the way, have partnered with plants in this manner since vegetation first colonized land. And a root microbiome well supplied with carbon from decaying organic matter and sugary exudates from plant roots is the key to ensuring that plants can acquire and take up an adequate supply of micronutrients and beneficial microbial metabolites (like those you read about in Chapter 3).

The next thing Lal wanted to show me was another experiment, this time comparing conventional fertilizer and compost. He'd started back in 1994 when he stripped all eight inches of topsoil off two side-by-side plots. He then began no-till cultivating the subsoil of both plots. He used chemical fertilizer on one plot and compost to deliver an identical amount of nitrogen to the other plot.

Guzman and Rimal waded into the cornstalks, pushed soil augers into the ground, and pulled up a sample from each plot. The chemically fertilized plot had an inch of light brown topsoil. The plot mulched with compost had six inches of dark brown topsoil above the khaki subsoil. New soil had formed six times as fast on the compost plot, at the rate of about a third of an inch a year. If you're not a geologist, that probably sounds slow. But consider how, at that pace, it would take less than a century to make a couple feet of topsoil. Other long-term studies from around the world have also found that manure, cover crops, and diverse cropping systems increase soil organic matter. Here was a recipe for how to restore topsoil and reverse historical soil degradation—surprisingly fast.

Like others I spoke with, Lal cautioned that, when implementing this approach, it takes two to three years to make the transition from conventional practices to productive low-input, no-till farming. It takes time to start building up soil organic matter, and this can present an intractable problem for highly capitalized farmers paying off big loans. Several years of low production can mean losing the farm.

CARBON SINK

Lal's plots were also demonstrating the largely untapped and under-appreciated role of soil as a reservoir to take up and hold carbon from the atmosphere.

The world's soils already hold at least twice the carbon as the atmosphere. Estimated to a ten-foot depth, soils contain more carbon than the combined amount in the atmosphere and all of the plant and animal life on Earth. Most soil carbon is held in the top several feet, due to surficial inputs of organic matter and the carbon-rich exudates that shallow roots push out into the soil. This means that changes in the organic-matter content of topsoil can significantly impact the amount of carbon in the atmosphere, and therefore global climate.

Every time we plow, it exposes soil to air, which speeds up decomposition of organic matter, releasing carbon skyward. Not just a little, but a lot. Since the start of mechanized agriculture, North America's tilled fields have lost more than 40 percent of their original soil organic matter. Before 1950, plowing on U.S. farms contributed more to our national carbon emissions than all other sources combined. By the close of the twentieth century, a quarter to a third of all the carbon added to the atmosphere since the Industrial Revolution came from

plowing. On the upside (I guess), a prominent climate scientist suggested that agriculturally driven loss of soil organic matter may have
delayed the return of another Ice Age.

Today, most croplands have been under conventional practices for
long enough to have reduced soil organic matter by more than half. In
a 1999 study, Rattan Lal estimated that the world's agricultural soils
had already lost 66 to 90 billion tons of carbon, mostly due to tillage
and the resultant erosion. He estimates that, since the dawn of agriculture, most cultivated soils have lost between a third and two-thirds
of their original soil carbon. It would take fundamental changes in
agricultural practices to restore soil carbon to near-historical levels.

But it could be done. Anthropogenic changes in soil organic matter are a two-way street—after all, people made terra preta. A 2005
assessment of the global consequences of land use in the journal *Science* reported that chemical fertilization and conventional cereal production decrease soil carbon by 0.5 to 1 percent a year, whereas the use
of cover crops and farmyard manure increase soil carbon by 0.2 to 0.4
percent a year. Estimates of how much carbon conservation agriculture
can sequester in different environments range from 0.2 to 1.0 tons per
hectare per year. This really adds up when you consider applying it to
all the world's agricultural land. In 1998, Lal and a team of colleagues
conservatively estimated that adoption of conservation agriculture on
U.S. cropland could sequester enough carbon to offset emissions from
half of the cars in America.

In addition to building up soil carbon, no-till farming and conservation agriculture also have lower fossil-fuel emissions than conventional practices. A 2003 comparison of average fuel consumption found
no-till farming used a third less energy than plowing, due to fewer
passes of equipment over the fields. And adopting cover-cropping
along with no-till further reduces energy-intensive fertilizer use.

There are additional benefits to increasing soil organic matter,
through its effects on a host of physical, chemical, and biological properties that influence soil quality and health. Farming practices that

build up soil carbon boost microbial biomass and nutrient cycling, and improve soil structure, texture, and aggregate formation—all of which enhance soil fertility. The net result is that increasing soil organic matter increases crop yields.

It also increases drought resilience by improving soil moisture retention and water-holding capacity, both of which will become increasingly important to modern agriculture in a changing climate. Reducing the organic-matter content of soil from 4 percent to 1 percent, as happened on many farms over the twentieth century, decreases the soil's water-holding capacity by about half. So the historical loss of soil organic matter on farm fields undermined the ability of crops to weather droughts. Reversing this loss would double the available water-holding capacity of the soil, vastly improving drought tolerance. Many of the farmers I met while researching this book described how yields from no-till fields consistently beat conventional yields in dry years.

Refilling the largest terrestrial reservoir of carbon—soil—offers a way to both offset a portion of global CO_2 emissions and reduce the effects of climate change on crop yields. But by how much? Estimates of the amount of carbon that can be added to agricultural soils vary widely, and a global meta-analysis found that no-till practices consistently increased soil carbon stocks only where used in conjunction with practices that increased biomass production and multiple crops per year. The effect of no-till on soil carbon sequestration is highly dependent on the cropping system it is used in conjunction with— residue retention, cover-cropping, and crop rotation. Adoption of no-till alone may do little to increase soil carbon.

Rattan Lal conservatively estimates that worldwide adoption of all three elements of conservation agriculture could put enough carbon back into soils to offset 5 to 15 percent of global fossil-fuel emissions. He says his Brazilian colleagues tell him his numbers are far too low. Other estimates of the overall potential for conservation agricultural practices to reduce global carbon emissions range from 1.2 to 3.3 billion metric tons per year—up to a third of total emissions. And while that won't solve our global carbon problem, even the low end of those estimates offers a good start on the road to doing so.

A more liberal 2014 estimate from the Rodale Institute suggests that the carbon sequestration potential of soils is even greater. By extrapolating reported rates from studies of regenerative organic agriculture to all global cropland, Rodale researchers estimated that carbon sequestration could offset between a quarter and about half of global greenhouse gas emissions, based on those generated in 2012. Their report further claimed that applying the potential maximum annual sequestration through regenerative practices on all the world's grazing lands could offset 71 percent of global emissions. When I asked Kristine Nichols about the difference between her and Lal's estimates, she replied that he assumed that soil organic-matter levels cannot rise above levels found in native ecosystems. At Rodale, soil organic matter went from less than 2 percent to over 5 percent in thirty years. They are now about back to their native soil baseline. By Lal's accounting, they'd be done. But Nichols thinks that they can push soil carbon higher with cover crops and manure—a lot higher.

Although estimates vary, agriculture is directly responsible for somewhere around 15 percent of global greenhouse gas emissions. So drastically cutting agricultural fertilizer and diesel use by adopting conservation agriculture practices would cut the agricultural sector's contribution to fossil fuel emissions by 5 to perhaps 10 percent. Add that to Lal's conservative estimate of 5 to 15 percent sequestration potential from the same practices and we can crudely estimate a potential to reduce or offset 10 to 20 percent of global carbon emissions. In addition, estimates of the additional carbon sequestration potential of biochar range from about 10 percent to well over half of global emissions. Again, even at these low-end estimates, there is substantial potential for soil-building practices to sequester enough carbon to make a real difference—if we can act over large acreages.

What will it take to build up carbon levels in degraded agricultural soils? Top-down soil building. The old bottom-up view of soil formation as a simple product of rock weathering held that it takes hundreds of years, or longer, to form an inch of soil. The new view that incorporates biology—cover crops, exudates, microbes and soil life—holds that we can grow soil much faster, in decades instead of centuries.

The majority of the carbon that builds up in the soil comes from root exudates. Tie up carbon in life forms and it can't ascend skyward. This is why mob grazing works so well for building up soil carbon—it stimulates exudate production like nothing else.

All this points to a big problem with conventional grain monocultures. As a crop begins to grow, it takes time before the plants build up enough photosynthetic carbon to produce exudates. And when the crop becomes reproductive, root exudates shut off as the plant shunts resources into seed production. So there is only a four- to five-week period when grains push exudates into the soil. That just isn't enough time to pump out much carbon. So grain crops don't contribute a lot to building up soil carbon. To dramatically increase soil carbon, cover crops are needed to push exudates into the soil over more of the year.

However, the carbon storage potential of soils is not unlimited. The pace of carbon buildup ramps to a maximum in the decades after adoption of new practices and eventually hits a plateau, above which it gets harder to further increase carbon storage. In this sense, soils are like batteries that can only get so charged up before it becomes pointless to keep charging them. Still, agricultural soils can potentially store a substantial amount of carbon over the coming decades. Researchers have estimated that it would take at least fifty years for European soils to max out soil-carbon levels after adoption of conservation agriculture. This means that soil-building agriculture can help buy time to transition to other energy sources or means of sequestering carbon before soil carbon storage becomes saturated.

Unlike many technologies to mitigate fossil fuel emissions, soil-building practices can be implemented immediately at low cost. And because these practices already make economic sense for many, if not most, farms, they might be readily adopted—if known about and promoted. If this sounds like an all-too-rare win-win situation, that's because it is, or could be if we pursued it.

FINDING A BETTER WAY

Just how feasible is large-scale, soil-based carbon sequestration? Can it work on typical North American commodity crop farms without livestock? These were questions I'd come to Ohio to ask. Gabe Brown had told me that I *had* to go see David Brandt's farm in Carroll, Ohio. Actually, he told me I had to see his soil.

For the past forty years, Brandt has run a farm-scale experiment in long-term soil building. He didn't do it to explore the potential to sequester carbon on farms. He did it because he believes it's the right way to build fertile soil on his farm. Yet in retrospect he's shown how profitable changes in farming practices can deliver a hefty side benefit of carbon emission offsets.

After sitting in horrendous traffic on the drive from Columbus, Lal and I pulled into Brandt's farm down a long driveway between fields and headed to the outbuildings behind the house. Brandt welcomed us jovially. An ex-marine who has been no-tilling since he came back from Vietnam, he has huge hands and a round face, which was shaded by a baseball cap. In his white shirt and blue-jean overalls, he looked the part of genial grandpa farmer.

But Brandt is far from a typical farmer. An unconventional corn and soy producer who owns 160 acres and leases another 800 acres, he's become a well-known leader in the cover-crop revolution. Mostly, he's followed a corn/soybean/wheat/cover-crop rotation. He plants cover crops into the wheat stubble two to three weeks after harvest. And when the roots of cover crops like buckwheat die, they release acids that help solubilize phosphorus and other mineral nutrients. If you don't kill the cover crop, you don't get this additional input, so if the winter doesn't kill the cover crop first, then Brandt finishes it off with an herbicide or a crop roller like Jeff Moyer uses at Rodale. The idea is to not let the cover crop pull nitrogen and phosphorus from the soil to make seeds, but to time the growth of cover crops in such a way as to release the nutrients conserved in their biomass back to the soil for his next crop to take up.

It was late in the day, and Brandt was eager to show us his fields. So he ushered us over to his four-wheeler. His dog, a Weimaraner named Yankee, jumped in, happy to ride along to the fields.

Brandt started planting cover crops back in 1978, when he seeded cereal rye onto bare fields to control erosion. Now he grows cover crop communities, with up to ten different types at a time in a field. He likes to feed his soil a diverse diet—a mix of cover crops, mostly legumes and radishes, which he knocks down to rot where they grew so they can nourish the subsequent cash crop.

Our first stop was a field with a ten-way seed blend, planted to prepare the ground for a crop of corn. Yankee hopped down to scout for something to chase. To me the field looked like a diverse stand of wildflowers. When Lal asked him how much of his farm is under cover crops, Brandt replied, "By fall harvest, 100 percent is under cover crops." When I mused out loud something to the effect of how much the field looked like a salad bar for cows, Brandt replied that he does not have any cows. On this farm the livestock is all underground.

He drove us across a field he'd planted with cover crops after harvesting 96 bushels of wheat per acre, and Lal asked about the greener strips running lengthwise down the field. "That's where a malfunctioning urea spreader dropped some extra nitrogen onto certain rows," Brandt replied. Fortunately, the cover crops were capturing the excess, keeping it in the field for his next crop to use instead of letting it run off to pollute elsewhere, like it would from a plowed field.

Our next stop revealed an elegant experiment, a side-by-side comparison of Brandt's field with the neighboring 81-acre farm he'd just bought. His side of the old property line had been no-till for forty-four years, while the neighbor's side had a long history of conventional tillage. When Brandt first started farming the neighbor's field as a tenant, it had 0.25 percent organic matter. Now, after two years of no-till and cover crops, it's up to 1.1 percent, an increase of almost half a percent per year. His goal is to bring his newly acquired fields up to 5 percent within seven years.

Experience tells him that the investment will pay off. In Brandt's recently acquired fields, the cover crops were half as high as in his no-till field. Before moving on, he reached down and yanked up a three-inch-long radish, a clod of soil dangling from its thick taproot. The soil was light brown with solid, platy chunks.

He then herded us back onto the four-wheeler and drove us over to his forty-four-year no-till field, which had been planted with the same cover-crop mix at the same time. Here, he pulled up a fat ten-inch-long radish. The soil was dark brown, crumbly to the touch, and gave off a rich earthy smell.

Back in the 1970s, Brandt's field had started at less than half a percent carbon, just like his neighbor's. In the decades since he started growing cover crops, it's increased to 8.5 percent; the carbon content of the native soil in his woodlot is not quite 6 percent. He's managed to raise the organic-matter level in his fields *above* that of the native soil, just as Rodale's Kristine Nichols predicted was possible. "How high do you think your fields will go?" I asked him. "I'd be happy if it stopped, but I don't think it's going to," he replied.

He gives his radishes credit for much of this transformation. Radishes are rapidly becoming a favorite cover crop among farmers, as both experience and research show their beneficial effects on soil quality and crop health. Radishes provide effective suppression of winter annual weeds from fall through early spring, and they grow taproots that can extend down three feet in just two months. When radishes rot after being killed by winter frosts, the large vertical holes they leave improve soil infiltration and help break up compacted ground. They have also been shown to increase plant-available soil phosphorus around their root holes, and to be excellent scavengers of soil nitrogen after summer crops. When they decompose, they rapidly release nitrogen and other nutrients back to the soil. All of this really adds up.

An analysis of Brandt's land that had been planted with tubers, like radishes, found that each acre recycled 250 pounds of nitrogen, almost as much potassium, and 23 pounds of phosphorus.

Other plants in his crop mixes serve other purposes. Sunflowers, for example, excel at scavenging zinc from depth and bringing it into the topsoil. Where Brandt plants sunflowers, he sees the plant-available zinc levels rise in his soil tests. Hairy vetch and Austrian winter peas have added between 100 and 200 pounds of nitrogen per acre. Since he's been planting diverse cover crops, he's found that his plant-available soil phosphorus and potassium levels have climbed, even though he's greatly reduced his fertilizer application. He says his agronomist tells him that's not possible.

Farther down the old property line, Brandt showed us the contrast in this year's corn on opposite sides of his old fence line. To our right was the field he bought two years ago. We passed six-foot-tall corn that had been planted along with soybeans and received no herbicide. He thinks he'll make 140 bushels an acre out of the field, just under the county-average corn yield of 145 bushels an acre.

Then the corn steps down to just three or four feet tall. He said, "You see what's happened here? This is where I ran out of soybeans." From there on, the field had instead received 160 pounds of nitrogen fertilizer per acre as well as herbicide treatments. He thinks he'll be lucky to get 100 bushels an acre from this field. We dug into the hard soil with a shovel. It was light brown, dry, and platy.

"Now look at this," he said as he waded into a thicket of corn rising several feet above his head on the other side of the road—the side that he's no-tilled for over four decades. This field has not had any fertilizer or herbicide for two years and no fungicide or pesticides for nine years. I bent down and dug into the soil with my bare hand. It was dark brown, moist, and crumbly, with a lush smell.

Brandt expected to get 200 bushels an acre from this field. Only he was planning to do it with no chemical inputs, thanks to his soil-building practices. And this wasn't an unusual year. According to him, his corn and soybean yields are generally at least 20 percent above county average. He does even better in dry years.

I was particularly impressed with the way Brandt runs his own experiments. Like most farmers in the county, he had been pairing genetically modified (GM) corn and a lot of herbicide. But he's experimenting with whether he really needs either.

This year he was comparing varieties of organic corn with varieties of GM corn, planted in narrow quarter-mile-long strips. These were not small-scale research test plots. So far, the organic seeds that cost $105 a bag were "all doing pretty good." He'd expected to have cutworms in the organic plots but didn't. He was comparing that with non-GM, insecticide-treated seed that cost $160 a bag, and similarly treated GM corn that cost $336 a bag. At these prices, he said, he'd need to harvest 45, 60, and 90 bushels of corn, respectively, to simply break even for the three types. This meant that the GM seeds would have to be twice as productive as the organic ones to pay off.

He also had a soil-life study done two weeks after planting. It revealed that there was half as much soil life where he'd planted insecticide-treated seed. "Did seed corn treatment eliminate my underground critters?" he asked. "Are we hurting ourselves using this in our fields?" I suspect that such curiosity underlies how he's worked out soil building practices that promote beneficial soil life.

He was also maintaining a control plot for his county's typical practice of full tillage and the addition of 200 pounds of nitrogen, even more phosphorus-rich row starter, and 2.5 quarts of Roundup an acre. With all these inputs, he'd invested $502 an acre. He predicts it won't make 100-bushel corn. That pencils out to a net loss at anything under $5 a bushel. Lately corn has sold for less than $4. With that system, the more acres he plants, the more money he loses.

In contrast, on each acre of his forty-four-year no-till cover-cropped fields, Brandt spent $137 on seed corn and used just 24 pounds of nitrogen at $0.60 per pound and an $8 quart of Roundup. Along with $160 cash rent, this comes to a total expense of just under $320 an acre. At the current $3.80 a bushel, his estimated corn harvest of 182 bushels an acre would return just over $690 an acre, a net profit of about $370 an acre, not counting his time, cover crop seed, and equipment costs.

Plus, Brandt saved more on diesel too. He says he used a bit more than 2 gallons an acre while, on average, his conventional tillage neighbors use more than 10 gallons an acre.

So while Brandt was using less than half the herbicide, about a fifth of the diesel, and a tenth of the fertilizer, he consistently outyielded his neighbors who practice conventional tillage with no cover crops. Unlike too many of their farms, Brandt's operation is actually profitable—quite profitable.

Consider, for example, the contrast with the farmer right across the road who recently bought 720 acres for almost $5 million. The farm had been run for thirty years with a corn-soybean rotation and no cover crops, wearing the soil down to just a few percent organic matter. The first year, the new owner used more than 220 pounds of nitrogen an acre and sprayed herbicides three times. Brandt estimated that "he's probably got $560 an acre in expenses and $320 for debt on the land. If he does really well and grows two hundred bushels an acre, he'll make $760 an acre. So he's already lost $120 an acre before he's even paid his diesel bill. How will he be able to keep going?"

Here is the trap that input-intensive farmers fall into, where they pay high costs at the front end and focus on yields and gross returns rather than on the spread between expense and income. High input costs and low commodity prices are a recipe for farm failure. This has been the story of the American family farm since the Second World War.

On his farm, Brandt has found that he doesn't need to add any nitrogen to a field once it reaches 8 percent organic matter. At that level, his cover-crop-fueled nutrient recycling is equivalent to applying 240 pounds of nitrogen fertilizer an acre annually. His input costs are low, yields are good even in dry years, and pests are not a routine problem or threat to crops. This sounded like resilient and sustainable farming to me.

The next day, Brandt picked me up at my hotel in his red Dodge Ram, a practical, no-frills pickup truck. The drive to the Ohio No-Till Council's field day workshop gave him a chance to express his concerns about how conventional agriculture treats the soil as a giant growing medium, with the expectation that if you just stick chemical nutrients in it, crops will come out. He wants everybody to learn to take care of the soil, because he fears that our society won't exist in a hundred years if we don't.

Experience is a good teacher and Brandt is a tinkerer—always testing out new ways of doing things, curious about what might work better. He likes to challenge guys who supposedly know it all. He laughed hard when I confided to him that universities are filled with such folk. When he first heard academics talk about an eight-way cover-crop mix, he was a single-species guy and simply believed an eight-species mix was a recipe for weeds. Still, he tried it out. After a year he was surprised to see positive changes in the soil and began experimenting with cover-crop mixes.

Brandt now sees cover crops as the key to supporting his microbial livestock. Cover crops don't just feed the microbes, they help moderate soil temperature so the microbes can work for him. In the hottest summer months, his neighbors' bare soils can heat to well over 110°F. But his soils don't get over 97°F. This matters, remember, because microbial activity pretty much stops above 100°F.

A decade ago he didn't know that there was a herd of livestock in the soil. He was taught that if he just applied the right nutrients he'd see higher yields. "I didn't know that earthworms eat our weed seeds, drag them down in winter, and manage weeds for us." Since then he's learned to feed his belowground livestock.

Brandt has spent his whole life on farms, the first of which was just a few miles from where he now farms. He turned seventy a couple months after I visited, and had started no-tilling in 1971—not by choice but out of necessity.

The day after he got married in 1966, he got a notice from his draft board to get a physical, and the next thing he knew, he was a sergeant in the Marine Corps. To this day he peppers his speech with "yes sir" and "no sir," a habit, I suspect, left over from those days. After serving in Vietnam for a few years, he returned home to find that his father had died without a will and his family had been forced to sell the farm.

His luck began to change when a niece of J. C. Penney (yes, that J. C. Penney) called, looking for a tenant to farm 400 acres under the guidance of an elderly farm manager who had worked for the Eisenhower administration. The manager told Brandt that the only way he'd be able to work the farm by himself was to go no-till.

When Brandt sold his tillage equipment and bought a no-till planter, his eighty-six-year-old grandfather, who'd plowed behind a pair of mules in his youth, was more than a little concerned. At the sight of his grandson planting his first no-till field, he took off his big straw hat and exclaimed, "My god, boy, what are you doing?" That year the corn made 127 bushels an acre. Grandpa was impressed. So was Brandt.

He did well at first but found that under his single-crop, low-residue routine, the ground became harder and yields starting falling. Even so, he refused to go back to tillage. So, in 1978, he planted hairy vetch and varieties of clover as cover crops. The hairy vetch worked so well that he stuck with it, experimenting with single-species stands of cover crops for twenty years. Then, in 1997, he decided to give peas and radishes a try, planting them in alternating rows. He immediately saw further improvement in his soil. After attending a field day demonstration at a conference in 2001, in which five-, six-, and seven-way cover-crop mixes were featured, he decided to try multiple species cover cropping. He was soon sold on the soil-improving power of combining no-till with a diverse mix of cover crops.

Brandt points to diversity as the key to success. Sunn hemp does really well with no rain while the hairy vetch will take off if it does rain. With a diverse cover-crop mix, he can be sure that something will always do well and take up any excess nitrogen left over from fertilizing. He's even found that if he starts his rotation with cover crops, he won't suffer any yield loss, even in the first several years after converting a field.

Particularly impressive was how, with nitrogen-fixing cover crops, he could cut nitrogen use in half after the first year. He liked the savings and recognized that the pollution implications of this are huge, as about half the nitrogen fertilizer U.S. farmers add to their soil isn't taken up by crops and runs off to cause problems elsewhere. Shortly before my visit, a health advisory alert for babies, pregnant women, and the elderly had been posted in nearby Columbus, due to the high levels of atrazine and nitrates in the city's drinking water supply.

When I asked Brandt if he'd like a bigger farm, he chuckled. "We can do more than plant corn and beans, but going bigger is not better." In fact, he wished that he could take 10,000-acre farms and break them into smaller farms, to be run by young up-and-coming farmers. Like Gabe Brown, he sees smaller, more profitable farms—more people on the land—as a solution to many of modern agriculture's problems. Amid the corn and soybean sea of Ohio's low hills, small towns with streets of neatly kept white two-story houses struggle to survive. It's easy here to imagine how reviving small farms could help revitalize small-town America.

"Who," I asked, "is teaching farmers to farm your way?" Brandt shook his head and drove without speaking for a while. Finally, he told me that he volunteers with Future Farmers of America and spends a couple weeks each year helping teach local kids the basics of his farming methods. He thinks that the adoption of conservation agriculture will depend on how we instruct our youth. And the United States needs some kind of program to connect young farmers to older farmers who are looking for someone to eventually take over their farms.

Like most farmers I talked with, Brandt was frustrated with most university researchers. "We can't get a university that is interested in doing research on no-till to look at cover-crop varieties. So we've got to do our own experiments, because most researchers only look at corn and beans." Of course, part of the lack of academic interest is that there is little funding for such research. Agrochemical companies aren't about to fund research on reducing the use of their products. And government-sponsored research tends to favor commercial partnerships over investigating and disseminating evidence-based practices that farmers can adapt and use to reduce their reliance on and expenditures for inputs.

Brandt worries that there are few university researchers studying examples of farmer success in reducing their inputs and maintaining or increasing their yields. "In my opinion, the university is twenty years behind us." I'd heard similar complaints throughout my trips. He's convinced that improving soil health can solve agricultural problems, eliminate erosion and flash floods, reduce pesticide use, and curtail nitrate and phosphorous pollution. If everybody planted cover crops, there would be no more algal blooms. Columbus, Ohio, would have good drinking water year-round, every year.

When I asked him about retirement, Brandt said that he's "not planning on it. Having too much fun." Besides, there are a lot of farmers yet to convert. He volunteered to demonstrate this fact by taking me around to compare his fields with neighboring conventional ones. There were plenty to choose from.

We arrived at a neighbor's conventional field and pulled off on the shoulder by a sickly-looking stand of knee-high yellowing soybeans. A quarter to a third of the field was covered with weeds, herbicide-resistant mare's tail and giant ragweed sticking up through the beans. I could see right through the vegetation to the dry, cracked ground beneath.

Brandt told me that it had been planted on May 1, four months ago, and repeatedly sprayed with Roundup since then. Looking it over, Brandt employed half a century of experience to confidently predict that it will yield 25 bushels an acre. "Ours are doing twice as well," he noted. We then drove a few hundred yards down the road to have a look at *his* soybeans.

Brandt's field was lush, dense with waist-high vibrantly green beans. I couldn't see the soil through the crop, but when I got out and walked into the field, I found the ground to be moist and loose. Last year this field produced 219 bushel-an-acre corn, and was then planted to rye. Brandt planted the soybeans now growing there on May 25, about three weeks *after* his neighbor had planted the sickly-looking yellow beans. Compared to his neighbor's conventionally sprayed field, Brandt's herbicide-free field was remarkably free of weeds. The only weeds I saw were the occasional volunteer cornstalk rising above the beans. And he didn't use any fertilizer on the field either—he didn't need to. "If we get more rain, my beans will put on more. But the neighbor's are done, they won't leaf out again and won't put more beans on even if it rains." He predicted that this field may produce close to 70 bushels an acre. Even I could tell that Brandt's plants held at least twice as many beans as his neighbor's.

On our way to the next field, Brandt mentioned that he has thirty-two separate fields in thirty-two separate locations. Is that hard to keep track of? "Yes." Does he ever forget a field? "Nope."

We stop next at a 90-acre field where he'd planted soybeans into rolled rye with no herbicide or fertilizer. The field was as lush as the one we'd just left. Again, I couldn't see the soil beneath the dense greenery, and the only weeds in sight were a few unplanned cornstalks rising here and there above the beans. Brandt had planted this field on May 26, and he estimated that it will produce 65 bushels an acre. Right across the road I could see the parched ground through the thin foliage of his neighbor's yellowing soybeans. Brandt said those beans *might* produce 25 bushels an acre.

As we arrived back at Brandt's farm, we passed a striking example of herbicide-resistant weeds. Across the road from the fields where he'd first showed Lal and me his radishes, about a third of the neighbor's field of yellow soybeans was covered with big green weeds—mare's tail. Brandt said that it had been sprayed three times already this year. It was not lost on me that almost all the weeds I'd seen were in herbicide-treated fields.

FROM GARDEN TO GLOBE

Brandt is eloquently plainspoken, with no stomach for bullshit. His genuine enthusiasm for soil health translates into trying to help others do what he's done. Although, as he says, "everybody's looking for a damn magic wand when there just isn't one," Brandt's confident that the principles behind the suite of practices he uses can be adapted to the conditions on any farm.

"Is the soil health movement catching on from the bottom up?" I asked him. "Yes, you're absolutely right," he said, something every academic loves to hear. "Most early converts were motivated by a financial crisis, but some are now starting to see good examples and opportunities and following them. And they are now going out and sharing with other farmers who see these aren't just some guys trying to sell them crap."

So, why don't Brandt's neighbors adopt his methods? Surely they must see the quality of his fields when they drive by. He, like many of the others I spoke to, blames crop insurance. Ensured a guaranteed return, farmers harvest healthy profits in good years and weather the bad ones with no incentive to adopt more resilient behavior.

Why, I wonder, aren't we framing agricultural policy to incentivize and make economically attractive practices that build soil organic matter instead of mining it? That question is finally starting to get some attention from farmers, academics, and government officials, in this country and others.

Brandt sees this in how he gets a lot of curious visitors. People drop by unannounced to see the guru of cover crops. Every now and then, a carload of farmers drives up to his place out of the blue. A scheduled busload arrives about once a month. He's in such demand as a speaker that he averages sixty to seventy days a year away from home. "Luckily we're no-till, so I have the time to do it." He is not at all proprietary about his ideas and practices, happy to invest the time he saves by no-tilling in telling others about how to transform their farms too.

Prior to my arrival, his most recent visitor was the French minister of Agriculture, Stéphane Le Foll, who has proposed a voluntary global effort to increase soil organic matter on agricultural land through the adoption of restorative soil-building practices, like conservation agriculture, mulch farming, cover cropping, and biochar. He calls his initiative "4 per 1000" because it aims to increase soil carbon by 0.4 percent per year. If achieved globally, this would sequester 2.8 billion metric tons of carbon annually. Add to that the reduction in agricultural fossil fuel use, and altogether it has the potential to offset a third of current global carbon emissions. That's enough to make a major impact on the climate problem by doing something that will help feed the world—and grow farmers' bank accounts.

When Le Foll explained his proposed initiative, Brandt told the French minister that 0.4 percent was aiming too low, explaining that, since adopting his cover-crop mix, he had raised his soil carbon content by more than half a percent per year. Le Foll all but high-fived him. When the minister asked Brandt to address a meeting in France, Brandt joked that, instead of wearing a jacket and tie, he would have to get formal black overalls.

Le Foll also invited Brandt to join him for lunch the next day at Ohio State University, along with four farmers and a crop insurance representative. Brandt sat silently as the others talked about how great the crop insurance program was, that they needed it to survive. When Le Foll asked him about his feelings on crop insurance, Brandt said that he'd carried it for two years in the 1980s when it was mandatory, but now he didn't need it. He hadn't had crop failures. Plus, he pointed out, if one of his crops did fail, there's always something else to plant that will grow fast. If you can't grow corn, you can plant sorghum. When you can't plant sorghum, plant canola. But those who have crop insurance will file a claim for the first failed crop and take the payment. "Crop insurance makes farmers lazy," Brandt concluded.

If they didn't have that insurance, they would be more interested in more resilient practices.

Aghast, the other farmers glared at the heretical lunatic. The farm bureau rep shrugged his shoulders and dismissively sniped, "But surely Mr. Brandt can't survive." Yet that's what he's been doing for forty-four years—he's the last of his high school class that's still farming. After the visit, the French minister's assistant contacted Brandt to let him know that he appreciated "learning about how important crop insurance is for U.S. farmers" in a way that made it clear that he had gotten Brandt's message.

Three months later, the French government proposed Le Foll's "4 per 1000" initiative at the 2015 U.N. Framework Convention on Climate Change in Paris. It was the first time in twenty-one years that the potential for changes in agricultural practices to increase soil carbon was discussed at their meetings. Unfortunately, it was not even mentioned in the final document.

It's not easy to quantify soil health, to reduce it to a single number. But the basic idea that more organic matter equates to better soil health is pretty robust. Every researcher and farmer I consulted offered the same answer—soil carbon—when asked what single metric they would choose to measure soil health if they could only use one. And it turns out that soil organic matter is pretty easy to measure. You can do it at home. The low-tech way is to get a sample of soil, weigh it, burn off the carbon in the oven, and weigh it again. What's been lost is the organic matter. The hard part is getting a meaningful average across a whole farm.

When my gardener wife embarked on the project of converting our barren yard into a garden, she didn't add any fertilizers to our soil. Instead, she focused on adding organic matter to fuel the microbial life that would attend to recycling nutrients and micronutrients and thereby support another round of life. Anne's plan for restoring fertility to our yard was a lot like conservation agriculture—minimal disturbance and mulching the soil. But how big a difference had it made to our soil?

At noon on a sunny day in late February (yes, they do happen occasionally in Seattle) my soil scientist colleague Sally Brown strolled up our driveway. I'd asked her to drop by to measure how much carbon Anne and I had parked in our yard. Sally studies how organic-matter amendments influence soil quality.

Her soil sampler consists of a six-inch-deep, two-inch-diameter metal cup with removable brass rings that line the sample barrel. The first step in sampling is to sweep away the loose layer of surficial organic matter from the ground surface. Sally wants the soil below. To collect each sample, she kneels and drives the sampler into the ground with its sliding hammer, a hunk of metal that slides up and down a metal pole. Once the sample barrel is flush with the ground surface, she digs around it with a shovel to pop the sampler out of the ground. Then she unscrews the sampler, removes and separates the three brass rings nested inside, and trims off the middle one with a deft swipe of a kitchen knife.

This gives her a sample with a known volume. Then she drops the sample into a paper bag and labels it with a black sharpie. She'll take it back to the lab and weigh it. From the weight of a sample of known volume, you can calculate the bulk density of the soil, its weight per unit volume. Then she'll burn the sample so that the organic matter combusts, and weigh it again to determine how much mass was lost—and thus how much organic matter was in the soil. "This is a measure that is simple, cheap, and tells you a ton," she told me.

At the time we bought the house, the ground was covered by a ratty century-old lawn, a wormless tangle of grass roots over lifeless khaki dirt. After all of our—well, really, Anne's—hard work, the contrast between the soil samples was obvious. Sally took samples from several places. Behind the garage, where we'd never mulched or planted anything, would serve as a control, a stand-in for our original soil. There the light brown soil cracked and arched around the sampler as she drove it into the ground. The ground cracked again when she used the shovel to unearth the sampler. She also took samples from a mulch-covered planting bed and from the lawn, where we allowed the worms to take grass trimmings into their burrows by simply leaving the vegetation on the surface after mowing. Finally, she sampled the veggie beds where Anne had used most of her kitchen-fed worm compost and occasionally a little biochar.

The sample from the lawn came out like a solid plug but was dark brown. The planting bed soil was looser and even darker brown. In the veggie beds, the soil was soft enough that Sally pushed the sampler in with just her hands, no need for the sliding hammer. It was blackish and loose, with a crumbly texture.

Several weeks later, she emailed us a spreadsheet with the results: big changes had happened in our yard over the years. The control sample came in at a little over 1 percent carbon, about the norm for degraded farmland soil. The lawn was at 4 percent, a significant increase that nonetheless paled next to the almost 9 percent soil carbon level of the mulched planting bed. A decade and a half of mulching and composting had dramatically raised soil carbon levels, bringing it up to levels comparable to David Brandt's forty-four-year no-till fields. And our veggie beds were approaching terra preta levels, with 15 percent soil carbon. Like Brandt's, our soil organic matter was increasing fast, at about 0.5 percent per year, in our mulched beds, faster than Minister Le Foll's target. The implication was as straightforward as it was profound. If everybody, farmers and city dwellers alike, increased the organic matter content of their soil along these lines, it could make a huge difference in terms of carbon storage, with wide-ranging benefits over the next half-century.

Soil carbon is not a plant nutrient, it doesn't feed plants directly and is not essential in and of itself to plant growth. Yet there is a close relationship between soil organic matter and crop yields. Increasing soil organic matter allows for maintaining the same level of crop yield with less fertilizer. In 2006, Rattan Lal published a paper that analyzed how yields for wheat, rice, and maize increased with the soil organic carbon content of the root zone. Based on the relationships for individual crops, he estimated that increasing soil organic carbon by one metric ton per hectare per year (a quarter to half a U.S. ton per acre) could increase grain production in developing countries by 24-to-39 million metric tons a year. This would be enough to meet three-quarters to more than the entire projected annual increase of 31 million metric tons of food production needed to feed developing countries in response to population rise and projected changes in diet in the coming decades. Efforts to feed the world would be well served by promoting agricultural practices that increase soil organic matter.

Way back in 1938, soil scientist William Albrecht wrote in the USDA *Yearbook of Agriculture* that "the maintenance of soil organic matter might well be considered a national responsibility."[30] We now live at a time when we need to update Albrecht's wisdom to *global* responsibility. We need to see soil carbon as a societal investment account—humanity's planetary nest egg. For, if we restore it now, our descendants can reap perpetual dividends. Unfortunately, we're still flushing it away.

12

CLOSING THE LOOP

All things come from earth, and all things end by becoming earth.

—Xenophanes

German chemist Justus von Liebig is heralded for his pioneering work on the efficacy of chemical fertilizers, like the effects of the big three known as NPK—nitrogen, phosphorous, and potassium—on plant growth. His seminal 1840 book, *Organic Chemistry in Its Application to Agriculture and Physiology*, discredited the then-popular humus theory of plant nutrition—the idea that plants directly ate decaying soil organic matter. This opened the door for substituting chemical fertilizers for traditional organic fertilizers like manure.

So when I read his 1863 book, *The Natural Laws of Husbandry*, I was surprised to find him bluntly contradicting the idea that soil fertility could be maintained by adding a few key substances to the soil. In this, his last major work, Liebig recommended returning organic matter to the fields to provide crops with a full complement of nutrients. It turns out that the patron saint of chemical fertilizers thought that soil organic matter *was* the key to sustaining civilization.

What had changed? Liebig came to realize that soil organic matter played a significant role in nourishing plants after all. While the carbon in decaying organic matter did not feed plants, organic matter had other elements plants needed. And adding an element or two through chemical fertilizers did not address the loss of these other essential nutrients. Liebig argued that farming practices that removed and didn't replace the full suite of mineral nutrients only worked for so long before they exhausted the soil.

Would Liebig really disapprove of our modern dependence on chemical fertilizers? Perhaps. Much to my surprise, he argued that it was difficult for chemical analyses to evaluate soil fertility because only a small fraction of the elements in soils were in a physical state readily accessible to plants. Fertility depended less on what chemicals made up the soil than on whether they were plant-available.

Liebig did not know about the role of soil life in mobilizing mineral elements and making them available to plants. But he recognized that mineral weathering proceeded slowly, and that elements tied up in rocks and soil minerals were not immediately available to plants. Manure and decaying plant matter, he recognized, provided accessible reservoirs of elements that could help maintain soil fertility. Continuously exporting crops from a field, without returning organic matter, led to the eventual degradation of the soil, he argued. And this had happened time and again throughout history. He pointed to the differences in agricultural practices between Western societies that exhausted their soil and those that farmed the same fields for millennia in the East. If agriculture were to last in the West it had to change.

He pointed to Chinese and Japanese agriculture that supported large populations for centuries. What was the secret to their success? These cultures routinely returned human and animal waste to the land. Liebig argued that the ability of farmyard manure to restore the fertility was established by "the experience of a thousand years."[31] He advocated for a large-scale return of organic matter to farms because "a soil deficient in organic matter must necessarily be less productive than a soil abounding in it."[32]

In his 1863 book, Liebig wrote that planting fodder plants to draw elements from soil particles, and place them into biological circulation, offered the best way to restore mineral nutrients to soils. Particularly effective were deep-rooted plants, like turnips, that drew elements up from the subsoil and returned them to the topsoil upon decay. Weeds too, he maintained, had value in nourishing plants on barren ground. For they would slowly extract and accumulate elements from the soil and, upon decay, return them in a plant-accessible state. In this manner, successive generations of weeds could enrich the soil, circulating mineral nutrients through cycles of growth and decay. This was how nature built fertile soils. Rotating crops offered a way to enrich arable land faster than nature would do on her own. Growing fodder crops like turnips and clover, and adding back the manure of the livestock that ate them, helped to renew fertility, prevent soil exhaustion, and ensure continued high yields.

Liebig was the last person I would have suspected to advocate regenerative practices. Yet what he recommended gets close to conservation agriculture—cover crops, crop rotations, and building up soil organic matter. He also encouraged farmers to conduct small-scale experiments to find out what worked best on their soil. Well aware of the variability of soils, he cautioned against applying the same techniques on all farms. He also warned against placing too much faith in simple chemical analyses of the soil, as farmers still do today.

Liebig compared continual withdrawl of minerals from a soil to withdrawing money from a bank for daily expenses without ever bothering to deposit more into the account. In his view, returning nutrients back to the land was best done through the excrement of animals (manure) and man (sewage).

As an example of the latter case, he described the Bavarian experience with poudrette, a fertilizer made from night soil. He related how troops garrisoned at the Radstadt fortress used privies constructed so that excrement fell into open casks fixed upon carts. This provided a convenient way to get rid of the waste. But Liebig noted that this solution had another advantage—it could sustain the fertility of the fields that fed the soldiers. Although dubious about accepting the stuff at first, peasants in the surrounding countryside lined up to buy dried night soil once they realized its fertilizing power.

In 1865, Parliament invited Liebig to visit England to advise on returning town sewage to farmers' fields. London's Great Stink, in the hot summer of 1858, had spurred the government to action as repulsed citizens blamed cholera outbreaks on the stench of untreated human waste discharged directly into the fetid River Thames. Liebig hoped that London would become an example for how to make use of all the plant food flowing every day from European cities as sewage. His idea had precedents extending back to Classical Greece, but that fact didn't make it any more popular. Instead, the appeal of indoor plumbing set society on a different course, in which cities began building sewer systems that ran to distant outfalls.

Liebig's goal of returning organic matter to the land fell prey to the success of the agrochemical philosophy his disciples had adopted. They focused on fertilizers as central to his Law of Compensation, as Liebig dubbed the need to replace the nutrients taken from a field in crops. Yet Liebig knew there could be no permanent agriculture so long as the full spectrum of nutrients were not returned. Decades later, Sir Albert Howard based his composting practices and his famous Law of Return on a similar argument about the need to return organic matter to the soil. Howard, however, offered his views as a direct refutation of relying on the fertilizers that Liebig's acolytes embraced. How interesting, I thought, that the fathers of fertilizers and organic agriculture both saw returning organic matter to the land as essential to maintaining fertility over the long run.

They were not the only ones. Franklin H. King's classic 1911 book, *Farmers of Forty Centuries*, argued that returning organic matter to the fields had supported intensive Asian agriculture for millennia—and was necessary to support any permanent agriculture. King has been called the father of soil physics, but he is perhaps best known for investigating why Asian fields remained fertile for thousands of years. As it turns out, his interest was partly of a personal nature—fueled by a bitter feud with his boss at the U.S. Department of Agriculture.

Raised on a farm in Wisconsin, King was interested in practical applications of knowledge. After attending Cornell University, he taught natural science for a decade to students training for teaching careers in River Falls, Wisconsin. Then, in 1888, King was appointed professor of agricultural physics at the University of Wisconsin. There he showed how switching grain silos from a rectangular shape to a cylindrical footprint reduced spoilage. This simple change in geometry eliminated hard-to-clean corners and reduced the potential to accumulate moldy grain. That such structures stood stronger against the high plains winds helped the design catch on rapidly—and persist to today.

In scientific circles King was known for his work on the movement of water through soils. Using glass columns packed with soil, some up to ten feet tall, he found that the size of soil particles influenced how fast water flowed through them. His interest in soil fertility was piqued when he learned about studies showing that dissolving potassium in soil water could increase crop yields.

In 1902, King was appointed chief of the Division of Soil Management at the USDA Bureau of Soils in Washington, D.C. In this new post, King's research began to show that crop yields correlated with the nutrient concentration in soil water rather than the bulk chemical makeup of the soil itself. Roots absorb dissolved nutrients in water, and what is in solution is but a tiny fraction of the total amount in the soil. And this, he concluded, meant that soil fertility was indeed exhaustible if farming practices depleted and did not replace the supply of plant-available nutrients.

This view conflicted with that of King's boss, bureau chief Milton Whitney, who believed that plant food was practically inexhaustible based on the bulk chemistry of soils. His view that the chemistry of soils was of marginal importance for crop production was based on finding little difference in the bulk amount of elements essential for plant nutrition in fertile and unproductive soils around Chesepeake Bay. Whitney maintained that soil fertility primarily reflected soil moisture and soil texture. Fertilizers, Whitney argued, boosted yields by affecting the physical properties of the soil. In his view, all soils contained enough plant food to continuously maintain crop production. This view was controversial, to say the least, when published as U.S. Bureau of Soils Bulletin 22 in the autumn of 1903. Whitney's argument was, in part, based on selective use of data from experiments conducted by King, who objected to his superior's biased interpretation and insisted that his name be removed as an author.

Frustrated by his boss both appropriating and, in his view, misrepresenting his work, King produced six manuscripts, three of which included data contradicting Whitney's conclusions in Bulletin 22. Whitney approved publication of three without endorsement, but he rejected Bureau publication of those that seriously undercut his own ideas. He then forced King to resign from the Bureau, driving both the man and his views from the young agency. In so doing, Whitney helped steer American agriculture down the agrochemical path.

In January 1904, King returned to Wisconsin. After publishing the papers Whitney had scratched, he became increasingly interested in how the farmers of East Asia had maintained the fertility of their fields for thousands of years.

His curiosity about Asian farming practices was rooted in concern over how to secure the future of his own country. King saw that the wealth of the United States came from a healthy supply of fertile soil relative to a small population. Preserving prosperity required conserving the fertility of the land, and by then it was no secret that American soils were degraded. Crop yields were propped up through importation and application of guano and mineral fertilizers. In the early 1800s, less than two centuries after European colonists first began to plow North American fields, the soils of the Southeast were so degraded that farmers surged across the Appalachians, drawn by the lure of fresh soils that had never seen a plow.

King, all too familiar with how American farming practices

degraded the fertility of the soil, believed that fertilizer use could only provide a temporary solution. Perhaps Asian practices held the key to establishing a permanent agriculture.

In 1909, he cashed in his life insurance policies in order to fund a nine-month excursion through Asia. On February 2, 1909, King, along with his wife Carrie, embarked on an epic journey to the Far East, seeking the secret to the longevity of the region's agriculture. Ironically, King's trip to the Orient happened the same year that Haber and Bosch developed their method to manufacture synthetic nitrogen fertilizers.

Across China, there was an average of almost 2,000 people, several hundred cows, and twice as many pigs per square mile. At such densities, a 40-acre farm that most American farmers regarded as too small for a single family would somehow support a village of 240 people, several dozen cows, and as many pigs. This penciled out to just a third of an acre per person of agricultural land. How was it possible that Asian farmers could do this, let alone do it for millennia?

On February 19, after weeks of gray skies and choppy seas, King and his wife landed in Japan's Yokohama harbor. The next morning, traveling to Tokyo, they passed a steady stream of men, horses, and oxen hauling loads of night soil back to the fields. Men pulled carts carrying six to ten tightly sealed wooden containers weighing 50 pounds each. King asked his interpreter why the city didn't discharge its sewage into rivers or a sewer system as a cheaper and more efficient means to deliver wastes to the sea. His interpreter replied that it would be wasteful to throw away such valuable material. Each year, Japanese farmers were spreading almost two tons per acre on average across the nation's cultivated land.

Human waste wasn't the only organic matter returning to Asian fields. During a visit to an orchard, King noticed how a covering of rice straw mulch kept the ground between the trees free of weeds. Japanese farmers also used straw as mulch for vegetable beds. The straw would be layered on top of the soil and then covered with a little earth, allowing water to infiltrate quickly and keeping it in the ground by reducing evaporation from the soil surface. As it broke down, the mulch also served as compost. Compost and mulch were the foundation of Japanese farming.

Estimating the amount of nitrogen, phosphorus, and potassium the Japanese returned to their fields, King found that they were adding back as much as they removed in their crops. They were closing the loop, cycling nutrients from the soil to crops to people, then back to the soil. Here was the key to sustaining agriculture over millennia.

King's ship reached Hong Kong on Sunday, March 7. As he explored the surrounding countryside, he noticed the painstaking way people gathered organic wastes and sent them off for composting. Ashes, night soil, and manure were all carefully collected, composted and then fed back to the crops. King was particularly impressed by how the well-laid stone floor of a pigpen had been washed clean, the effluent collected for use as plant food. Along a riverbank, he watched men empty a boat of its two-ton load of night soil, carefully dilute it with water, and spread it on beds of leeks. A pair of pails balanced across their shoulders, the men used long-handled dippers to ladle this brown gold onto their soil.

Across the Far East, King estimated that each year people sent back to their fields 182 million tons of human waste containing over a million tons of nitrogen, some 376,000 tons of potassium, and 150,000 tons of phosphorus. He contrasted the continuous need for enormous quantities of mineral fertilizers in American fields with the lack of mineral fertilizers in Asian agriculture. One system degraded soils and produced declining yields in a century or two. The other sustained high crop yields for millennia.

The cause of this difference was clear to King. Each year, the United States and Europe discharged into the sea some 5 to 12 pounds of nitrogen, 2 to 4 pounds of potassium, and 1 to 3 pounds of phosphorus for every citizen. Somehow, he pointedly remarked, this net loss of fertility was viewed as one of civilization's great achievements. In his view, the Asian system of farming better fit the bill.

The secret wasn't just closing the poop loop. Long experience had taught Asian farmers that including legumes in crop rotations bolstered soil fertility. Farmers would plant clover just before or immediately after harvesting a rice crop. Later, it was turned under or pulled up, dipped in mud from irrigation canals, mixed with soil and composted, or allowed to ferment for several weeks. But organic matter was always applied back to the fields.

It was common to see a subsequent crop planted in the spaces between rows of a crop approaching harvest, the new crop getting a head start before the old was harvested. King stopped at a one-acre cucumber field and inquired with the farmer about his yields. The man's human-manure-fertilized field produced four times the yield to be expected from American methods. In addition, each season the man's field also produced two crops of greens grown between the cucumbers. The total amount of food produced per acre was off the charts compared to American farms.

Japan's Bureau of Agriculture estimated that in 1908 almost 24 million tons of composted animal and human wastes were applied to the nation's fields. That same year, the City of Shanghai sold the right to haul away night soil from public parks and private residences for the princely sum of $31,000. Shanghai's health officer, Dr. Arthur Stanley, confirmed to King that bacterial degradation was an effective way to purify the composted fecal matter and household waste. To discharge such material into a river was to destroy a water supply. To compost it was to reclaim its nutritional value.

After taking a steamship upriver to the steep hill country of Shantung Province, King visited a prosperous rural village to see how they made compost. Tidy piles lay in front of each house in the village, as dozens of men along the main street mixed manure, household wastes, and crop stubble, stirring repeatedly to aerate the mixture. Periodically, water was applied to the compost piles to control the pace of fermentation. After the organic matter broke down, the compost was mixed with fresh soil and ashes, dried, and either pulverized by hand or crushed beneath stone rollers pulled by livestock. The finished material was transported back to the fields and applied before the next planting.

King was intrigued with the way Shantung farmers sowed cotton-seeds on the ground surface ten to fifteen days before the wheat was harvested. The top several inches of soil in the furrows between crop rows was then loosened, dug up, and used to cover the just-sown seeds. This earthen mulch helped keep the soil moist and promoted germination. No plow touched this deep, loose soil. When the wheat was harvested, the cotton plants were already pushing up through the mulch and stubble, having gained weeks of extra growth. Here was the key to controlling weeds: plant the next crop slightly before harvesting the previous one to give the crop a head start so it could elbow out the weeds.

King calculated that were his home state of Wisconsin to support the same density of human settlement as the most productive part of Shantung Province, it could hold 86 million people, as many pigs, and 21 million cows. This was nearly equal to the entire U.S. population at the time.

Staggering the growth of multiple crops in the same field was a common practice among Chinese farmers. King marveled over how a field could hold several crops in different stages of development. A crop of winter wheat nearing maturity could shelter a half-mature crop of beans and just-sprouting cotton. Crop rotations alternated legumes with wheat or cotton. Multiple cropping was the rule rather than the exception. As I read this, I was struck by the parallels with modern-day conservation agriculture.

At the Nara Experiment Station in Japan, King noted how soybeans were planted between rows of barley to provide green manure (and, as we now know, provide nitrogen to the barley). A week after the barley was harvested, the soybeans were harvested before planting a crop of rice. These farmers followed a four-year rotation of barley, then rice, a summer melon crop, and some other vegetable before returning to barley and rice again. In Korea, the farmers alternated two rows of corn or millet with a row of soybeans. King saw no good reason why American farmers could not likewise adopt crop rotations and inter-planting as well.

Yet King did not advocate simply copying Asian practices. He held that they offered lessons that could prove critical for Western agriculture. I found it notable that a keen observer like King never mentioned encountering any concerns over plant pathogens or insect pests from Asian farmers.

Tragically, King died on August 4, 1911, before finishing his book. His wife saw it through to private publication in Wisconsin, lacking the final chapter of conclusions King had planned. The book languished until 1927, when a London publisher picked it up and Sir Albert Howard subsequently cited it in his foundational works on organic agriculture. Then, in the 1940s, Jerome Rodale republished it. A little-read classic, King's book presciently identified foundational practices of Eastern agriculture that parallel the principles of conservation agriculture.

Modern efforts to bring organic matter from urban centers to farmland offer new takes on Liebig and King's vision of large-scale nutrient recycling. Municipalities around the country are increasingly returning composted food, household, and yard waste, as well as microbially digested human waste—biosolids—to build up soil fertility in forests, farms, and community gardens.

This is not a new idea. Milwaukee, Wisconsin, has been selling Milorganite—discretely packaged biosolids—to gardeners and farmers for almost a century. Denmark has a long tradition of land application of sewage sludge, recycling more than half of what the populace produces. Canada returns about a fifth of its sewage to the land and the U.K. returns twice as much. And it turns out that the same thing is happening just south of Seattle, where I live.

At 11 A.M. on a sunny August day during the hottest summer on record in Seattle, Sally Brown and I arrived at Dan Eberhardt's office at the City of Tacoma's sewage treatment plant. Eberhardt oversees the transformation of sewage into TAGRO, a microbially digested fertilizer wildly popular with the city's home gardeners. The mayor loves it too—it makes the city a load of money, recouping about half the plant's $1.5 million annual budget.

Eberhardt looks more like a guy who'd fix your motorcycle than a municipal alchemist who turns poop into gold. He has a silver-tinged goatee and thick tattoo-covered arms; three diamond studs grace his left ear, and a hoop earring dangles from the right. I suspect that he wouldn't mind if the large Harley-Davidson sign that dominates his office wall came across as a bit intimidating. But his big, reverberating laugh is warm and open. It immediately put me at ease.

Most wastewater treatment plants are closed to the public. Tacoma's not only encourages visitors; it encourages them not to leave empty-handed. The mayor promotes urban agriculture on vacant city lands and provides more than seventy community gardens with free TAGRO. This fertilizer is so popular that there's not enough to go around; last summer they sold out in June. Eberhardt calls TAGRO "the gateway drug to better gardening."

TAGRO is a blended mix of biosolids, sawdust, and sand. Sally describes it as the ultimate potting soil and calls this sewage treatment plant a soil factory. I came to think of it more like an enormous industrialized cow—a municipal rumen producing processed humanure.

As one might imagine, controversy and concern over the safety of biosolids like TAGRO tend to center on the presence of heavy metals, pathogens, pharmaceuticals, antibiotics, and personal care products in municipal sludge and wastewater. Eberhardt says that's not much of an issue for TAGRO. Most of the heavy metals like cadmium and chromium were eliminated as Tacoma's industrial base declined and moved overseas. And microbial digestion during processing takes care of pathogens and most pollutants. In the end, what comes out of the plant is primarily dead microbes, the original ones having been eaten over and over again. While there are obvious concerns about the wisdom of applying raw or contaminated sewage to the land, Eberhardt and Brown both say properly treated and applied biosolids are perfectly safe. So does the EPA, which deems TAGRO safe for gardeners and children's play areas.

Still, the city's wastewater lab regularly tests TAGRO and the soils at urban gardens where it is applied. Class A biosolids, like TAGRO, are tested for complete pathogen kill and heavy metals and must meet the set limits. If they do, they are unregulated and can be sold to home gardeners. Class B biosolids do not require complete pathogen kill, and are applied to forest and farmland or disposed of in landfills. When applied to the soil, microbes prey on pathogens and exposure to sunlight and oxygen helps kill pathogens adapted to dark, anaerobic environments—like inside of us.

Eberhardt led us from his office toward an open shed housing big black piles of TAGRO, located behind a sign advertising that people can take as much as they want for $10 a cubic yard. I watched as an elderly couple loaded several 5-gallon buckets into the trunk of their immaculate sedan. Then a guy in a baseball hat and red sleeveless shirt drove up, his tattooed arm hanging out of the window of his large white truck. "This is way too good for just ten bucks," he said as he shoveled TAGRO into the bed of his pickup. Eberhardt beamed as the guy gushed unsolicited testimonials. "This stuff is amazing. I put it on my lawn and my grass went nuts. I love it." That, Eberhardt says, is pretty much how it works—word of mouth. People who try it love what it does to their gardens and yards. They tell their friends and come back for more.

Eberhardt describes the biosolids his plant produces as "dead microorganisms—a bunch of dead bodies." That's central to why it makes it such good fertilizer. Bacteria in particular are rich in nitrogen and phosphorus, and biosolids that become TAGRO are about 6 percent nitrogen and 2 percent phosphorus. Once mixed with sawdust and sand, it drops down to 1 percent nitrogen. That's still a lot of nitrogen. The stuff is like steroids for plants.

On the short walk over to the wastewater plant, Brown related how municipalities are starting to see biosolids as a resource and to think of wastewater treatment plants as resource recovery plants. For the next hour, I got a private tour of the facility, something I never thought I'd find interesting, let alone fascinating.

As raw sewage arrives at the plant through four-foot-diameter pipes that never stop flowing, it's screened through a big metal grate that strains out all manner of things—like two-by-fours, cell phones, and kids' toys—that then end up in a dumpster. In the next step the grit (silt and fine sand) settles out into big, open-top collection tanks. Next, the scum, grease, and oil are skimmed off the surface, and the solid sludge and liquids are split to be processed separately. The solids are then heated up to pasteurize the sludge and kill off microorganisms and pathogens. After that, oxygen is pumped in to fuel the growth of microbes that aerobically digest the pasteurized sludge. After aerobic digestion, the sludge goes through a series of methane digesters. Here anaerobic bacteria gorge on the bodies of their dead brethren that grew in the aerobic tank. This produces methane gas that helps power the heat exchangers and run the treatment plant. The combination of aerobic and anaerobic digestion breaks down most organic contaminants and ensures little chance of anything nasty making it through alive.

Whatever you may think of biosolids, Tacoma's treatment plant really does have a good business model. As Brown and I headed back to my car, we passed more people loading up their trunks and trucks, buying back what they'd paid the city to take away in the first place.

NOURISHING OR NOXIOUS?

TAGRO may be the gold standard in biosolids, but not everybody likes the idea of applying biosolids back to the land. Questions and concerns remain, even though studies generally don't report evidence of serious problems. Those that do find a wide range of chemicals in biosolids generally find them at low concentrations. Yet not everything found in biosolids is tested for and tracked, and some organic chemicals won't break down and can accumulate—like phthalates, plasticizers, and surfactants. The effects of exposure over time, or potential synergies with other toxins that a person encounters in their lifetime are among the issues that concern detractors.

In addition to what's in biosolids, there are concerns over how and where they get applied back to the land. I've heard stories about hikers finding biosolids sprayed several feet thick on forestland in Washington. And controversy rages over biosolids applications in British Columbia and elsewhere where complaints of adverse health affects have arisen after repeated heavy applications.

There was widespread concern back in the 1970s and 1980s over applying raw sewage to land, particularly in regard to heavy metals and pathogens. But source control activities greatly reduced concentrations of heavy metals in biosolids. Portland, Oregon, for example, reduced them by two- to tenfold from 1981 to 2005. Although all kinds of pharmaceutical compounds are found in sewage, the combination of anaerobic digestion and aerobic composting with the right biology breaks them down—most of them anyway.

Still, the concentration of contaminants in soil resulting from land application of biosolids will reflect the concentration in the source sewage and how the biosolids are made, as well as the rate of application and local farming practices. Of course, using raw, poorly made, or contaminated biosolids may lead to serious health affects. But published studies have yet to document strong evidence of serious health risks from well-made biosolids. For example, a 2014 analysis by Italian researchers found the risks associated with land application of biosolids to be very low in terms of pharmaceutical and personal care products. Biosolids proponents point to a 2002 National Research Council report on biosolids that found no documented scientific evidence for agricultural use of biosolids having an adverse impact on public health. Critics point to the same NRC report as recommending further studies to evaluate standards, document contaminant and pathogen levels, and develop risk-based assessments.

Despite vocal opposition, biosolids production has not slowed. In 2014, California made 688,000 dry metric tons of biosolids, about a tenth of national production. Chicago makes 200,000 tons of biosolids a year and now allows use of it on vegetables. Biosolids have been used to help reduce erosion, improve water quality, and restore soil health on fire-damaged land. It has also been used to reclaim soils on superfund sites, mines, brownfields, and overgrazed rangelands. About half of the biosolids produced in the United States are applied to the land, mostly for agriculture. The other half are incinerated or dumped in landfills. Even so, biosolids are applied to less than 1 percent of U.S. farmland.

There is no controversy, however, over whether the stuff makes great fertilizer. Biosolids are mostly water (60–80 percent) and carbon (4–10 percent), but they are also rich in nitrogen (2–6 percent), phosphorus (1–3 percent), and sulfur (1–3 percent). In addition, they contain trace amounts of more than eighteen micronutrients that plants need, including boron, calcium, chlorine, cobalt, copper, iron, magnesium, manganese, and zinc. In 2015, the net fertilizer value of the nitrogen and phosphorus alone in northwest biosolids was more than $40 a dry ton. Spreading several tons per acre can provide enough nitrogen and phosophorus to fertilize even the most degraded soil. Most of the nitrogen is tied up in organic matter and slowly becomes plant-available. This means that, unlike soluble chemical fertilizers, the nitrogen in biosolids sticks around and can build up.

One study from Washington State University found no difference in crop yields between dryland fields in eastern Washington that had been fertilized with chemical fertilizers and those fertilized with biosolids. All of these fields had lost half their soil carbon in a century of conventional farming. But the soil carbon content increased 0.4 to 0.5 percent per year in fields fertilized with biosolids—right on target for Minister Le Foll's initiative. Other published, long-term studies of biosolids application in Washington, Illinois, and Virginia likewise reported significant increases in the carbon content of soils amended with biosolids. Long-term studies have found that both compost and biosolids applications also increase soil nitrogen and phosphorus content, soil microbial biomass, and water-holding capacity. In other words, contaminant issues aside, biosolids can help rebuild soil fertility.

But even biosolids proponents estimate that there is only enough supply nationally to fertilize a fraction of the nation's arable land. So consider the potential to use biosolids and composted food waste as the ultimate locally sourced fertilizer for urban agriculture, assuming, of course, proper care has been taken to reduce contaminants in the waste stream and to ensure thorough aerobic and anaerobic microbial digestion and composting.

Urban farms can efficiently recycle urban organic wastes—food waste, garden waste, and even human waste—close to where they are produced. This is not a radical new idea. Urban farming supplied nineteenth-century Paris and twentieth-century Havana with fresh vegetables. London helped feed itself this way during the Second World War, as did Victory Gardens in the United States, which grew almost half the vegetables Americans ate during the war. Today, urban food gardens are once again gaining in popularity, with waiting lists for U.K. allotments, Swiss Schreber gardens, and American P-Patches. Highly productive farms are springing up in the heart of cities around the United States, from Los Angeles to Detroit. As of 2009, humanity is officially an urban species, with more than half of us now living in cities. It will become increasingly challenging to feed large urban populations as we pave over farmland ringing cities and as the costs of transportation from distant sources rise along with energy prices.

A key reason to consider returning organic matter to the land is that we do not have an inexhaustible supply of mineral fertilizers, phosphorus in particular. Crops need a lot of this element relative to what most rocks contain, and mineral deposits are not uniformly distributed. And because we use the highest-quality rock first, the quality of what's left dwindles as we use more. U.S. reserves are declining rapidly, and China now accounts for almost half the world's phosphorus production. Morocco sits on three-quarters of the world's phosphate rock. Yet demand for phosphate fertilizer is projected to increase in the coming decades. At the same time, the excess phosphorus that now runs off from farms, feedlots, and wastewater treatment plants damages water quality and aquatic ecosystems.

While it makes sense to recycle organic wastes to help meet our need for phosphorus and protect the environment, biosolids alone won't solve the problem. A recent study found that recycling human urine and feces could meet less than a quarter of the global demand for phosphorus.

But we could get a lot more mileage out of animal manure. A 2012 analysis found that U.S. livestock excreted enough phosphorus to make up for more than 85 percent of what crops remove, if it could all be recaptured. But today, few farms have much livestock, and manure is not evenly distributed. Instead, it accumulates in enormous brown cesspools at confined animal-feeding operations. A 2011 study of the annual phosphorus budget of England found that it would take hauling 2.8 million tons of manure across the country to balance phosphorus supply and demand. An efficient way to do this, of course, would be to co-locate livestock and fields and let cattle graze crop stubble, something better suited to their biology than eating grain in feedlots.

In the United States, a 2016 study examined the potential for phosphorus recycling using 2002 population and agricultural census data to account for the production of animal manure, human excreta, and food waste on a county-by-county basis. The international team of researchers then compared the amount of potentially recoverable phosphorus to the amount used to fertilize corn. Surprisingly, these three sources could supply more than two and a half times the annual demand for phosphorus fertilizer in American cornfields. Manure accounts for 90 percent of the supply, and three-quarters of the overall demand could be satisfied from within the same county. Of course, doing this would bring up other issues, like how veterinary pharmaceuticals, especially antibiotics, have been reported in soils as a result of animal manure applications.

Still, we could fully meet our national demand for phosphorus fertilizers for corn through recycling just over a third of the manure our livestock produce. Yet, at present, only 5 percent of U.S. cropland receives any manure at all. Instead, we flush phosphorus into rivers, lakes, and seas. An analysis of the phosphorus budget for Minneapolis-St. Paul found that 96 percent of what entered the city was lost as waste. If recovered, it could fertilize enough land to grow half the city's food supply.

ENDLESS RENEWAL

Fulfilling Liebig's final vision would take a farming system reconceived from the ground up in terms of how we work the land, what we grow, and what we do with organic wastes. At a national scale, this system would involve embracing soil-building practices in rural areas, and recycling urban organic wastes to return organic matter back to farms—urban or rural.

Take it one step further and imagine the potential to restructure our agricultural system around crop rotations to produce the four F's of food, fiber, fuel, and fodder. If the basic ingredients for industrial production are biodegradable, then societal waste products—food scraps, biosolids, clothing, and consumer goods—can all get composted and returned to the fields as fertilizer. Biodegradable plastics already exist, made from corn biomass. Clothing and consumer goods made from plant fibers could be composted and returned to the land. Particular mixes of crops would vary with regional climate and soil type, and reflect integration of food, materials, energy, and clothing in diversified crop rotations. This could include food crops like corn, wheat, and soy along with cover crops that bolster soil fertility and serve as livestock fodder, textile fibers like cotton, hemp, and flax (linen) for clothing, and canola, sunflower, and other quick-growing biomass crops for oils and fuel (biodiesel). Imagine trains coming into cities full of food, fiber, and fuel, and leaving full of compost, biochar, and biosolids, returning to the fields what we took from the land in an endless cycle flowing into cities and back out to the fields.

Is this a utopian fantasy or a prophetic vision? Maybe both. To achieve such a vision we'll need to develop incentives, markets, and infrastructure so that farmers can incorporate these crops and practices into commercial rotations. Yet this could be a blueprint for a diverse, resilient agriculture capable of maintaining soil health and supporting dense human populations.

The foundation for it all is soil-building. And a proven recipe for that is to increase soil organic matter. We know it works. It's the same recipe for making terra preta. So imagine a modern initiative to push the recycling of urban organic wastes—a century from now we could have rich fertile land to grow greens, fruits, and vegetables in and around cities, thereby slashing the distance food travels from farm to plate. And if we were to anaerobically digest or char organic waste in cities, we could produce energy to ship char and compost back to agricultural land.

Each year, the average person in the United State generates almost 50 pounds of biosolids, 175 pounds of food waste, and 200 pounds of yard trimmings. About half of the biosolids and yard waste end up in landfills. Almost all of the food waste does too. If properly processed, composted, or charred, most of this material can be recycled and put to use restoring fertility to degraded farms or building urban soils. Instead, we treat it like dirt.

It is understandable that there is resistance to the idea of putting sewage, even thoroughly processed stuff, back to work on farms. Still, it's not crazy to think that over the next hundred years we could safely reengineer our waste stream in ways that may seem unimaginable today. Recall that just a century and a half ago there was no indoor plumbing.

What would Liebig think of our shortsightedness? I suspect he'd say that we have a systemic problem with nutrient transfers. Instead of closing the loop, we add fertilizer to the fields and manure doesn't go back to where it came from because cows aren't in the fields, and there's no money in hauling crap. It's a great system—if you happen to be selling fertilizer.

13

THE FIFTH REVOLUTION

The biggest barrier to agricultural progress is between the ears.
—Kristine Nichols

As I started working on what would become this book, my wife, Anne, and I visited the Eden Project in southern England—the world's largest greenhouse, made of enormous geodesic domes built in the open pit of an abandoned clay mine. One exhibit featured a twenty-foot-tall nutcracker designed by an industrial absurdist. Scrap-iron pulleys, chains, cranks, and levers launched big metal marbles down tracks to turn gears that slowly hoisted a wrecking ball into the air before dropping it onto a carefully positioned hazelnut. Kids competed to power the thing, turning a crank on the side of the enclosure that held the device. We joined an enthralled crowd and watched the intricate dance of parts designed to solve the problem of cracking a nut.

Upon leaving the pavilion, Anne pointed out plenty of perfectly good rocks lying around that could do the same job with little effort in a fraction of the time. Here was the exhibit's broader lesson. Even with simple solutions in plain sight, complex ones attract our attention—and interest.

But simple ideas that solve problems do catch on. And visiting farmers around the world who were doing well putting regenerative agriculture into practice convinced me that building soil health offers a practical, cost-effective way to restore degraded land and maintain or increase crop yields with less oil and agrochemicals. Seeing how these innovative farmers restored their soil, their farms, and their bank accounts convinced me that we could avoid the fate of past civilizations. It's not a question of if we can, but whether we will.

Conventional wisdom says that fertile soil is not renewable, that it can't be replaced. But that's not really true. Fertility can be improved quickly through cover cropping and returning organic matter to the land. Soil-building is about getting the biology, mineral availability, *and* organic-matter balance right, rolling with the wheel of life instead of losing ground pushing against it.

As we've seen, restoring fertility to the world's cropland is not an either-or choice between modern technology and time-tested traditions. We can update traditional wisdom *and* adopt new agronomic science and technology. Solving the problem of land degradation is devilishly simple from a practices standpoint. The difficulty lies in marshaling the political wherewithal to stop subsidizing conventional farming and start promoting practices that build soil fertility.

The principles of conservation agriculture offer flexible, adaptable guidelines for restoring soil health, feeding the future, and ensuring that farmers can make a living without damaging the environment. Everywhere I went, from the tropics to the plains, I found that farmers who minimized soil disturbance and adopted practices to increase soil organic matter and cultivate beneficial microbes could build fertile soil on both conventional and organic farms.

Of course, the specifics vary. Every farm is unique to some degree. What works well in temperate grasslands may not work so well in tropical forests. We need to tailor practices to the land and be mindful of geographic and social context, as we seek to optimize the use of land, labor, chemical and organic inputs, and machinery to increase farm profitability *and* soil health.

The way to meet this challenge is to figure out how to get farmers to adopt practices covering all three principles that work for them on their farms. For it does take all three—minimal soil disturbance, growing cover crops, and devising complex rotations that work together as a system. Leave one piece out and it doesn't do what it's supposed to, just like a stool that needs all three legs to stay upright.

The farmers I visited are not pushing ideas to sell other farmers anything, land their next grant, fatten their reelection coffers, or please a funder or employer. They share a deep sense of community and want to pass on knowledge of a system that works well for them—and could for others too. And while they came to this viewpoint via their own experiences, they are not alone.

Both the United Nations Food and Agriculture Organization (FAO) and the World Bank recommend the three elements of conservation agriculture as the key to sustainable development for small farms in the developing world. The World Bank promotes these same principles as the basis for "climate-smart" agriculture to increase crop yields, reduce greenhouse gas emissions, sequester carbon in soils, and bolster agricultural resilience to climate change. Even agrochemical giant Monsanto now advertises soil health as central to the future of agriculture. If organizations across the ideological, political, and industrial spectrum agree on the need to adopt practices that enhance soil health, why aren't we promoting this with all the tools in society's policy toolkit?

Such a fundamental realignment of agriculture means big change across the board. There will be supporters and resisters. Who has the most to lose? Those who make and sell the agrochemical inputs on which conventional agriculture now relies.

Curiously, many of the arguments about conventional versus organic agriculture break down when viewed through the lens of soil health. Organic farms that adopt practices to boost soil health become more productive and conventional farms become more profitable. Recent reviews of nutritional studies report that organic foods not only have lower pesticide residue but higher phytochemical, antioxidant, and micronutrient density as well. What if we could get these health benefits through minimal fertilizer and pesticide use, without going completely organic? Conservation agriculture offers such a possibility.

Converting conventional farms to lower-input practices also would help address problems of soil erosion, water retention, energy use, and nitrate, phosphate, and pesticide pollution. If improved soil health became a consequence of agricultural production, this would not only solve agriculture's oldest problem but help address some of the most pressing issues humanity now faces.

For soil restoration offers a triple harvest of societal benefits, along with better farm profitability. It simultaneously builds soil fertility to help feed the world and improve food quality, stores carbon to slow climate change and boost agricultural resilience to it, and conserves biodiversity on agricultural land. As a bonus, taxpayers could save money through reduced subsidies.

Restoring fertility to the world's degraded agricultural soils would reduce our dependence on energy-intensive practices and help sustain high crop yields in a postoil world. The farms I visited showed that yields under fully established conservation agriculture systems can meet or exceed those from conventional agriculture. And while the transition may take several years to pencil out, it makes far more sense over the long run.

A 2006 assessment of low-input, resource-conserving agricultural practices in 57 countries in Latin America, Africa, and Asia evaluated 286 development projects that used cover crops for nitrogen fixation and erosion control, applied pesticides only when crop diversity and rotations were not effective for pest management, and integrated livestock into farming systems. For a wide variety of systems and crops, the mean increase in yields was 79 percent, not quite a doubling of harvests but enough to feed the world of tomorrow if achieved globally. For projects that had data on pesticide use, yields grew by 42 percent, while pesticide use declined 71 percent. Many of these changes were attributed to practices that improved soil and crop health, and thereby allowed effective pest control with minimal pesticide use. This is evidence that more diversified, low-input farming can work for many subsistence farmers.

As a general rule, ecologists find that systems with greater diversity are more resilient. Monocultures rarely exist in nature. If they do arise, ecosystems with a single dominant organism don't tend to persist. On farms they are just as unstable and vulnerable to pests and pathogens. In constract, greater on-farm biodiversity is a recipe for resilience against pests and pathogens that's been field-tested in nature for hundreds of millions of years.

We have rules and regulations to prevent industries from polluting rivers and streams. Farmers shouldn't be allowed to either. No one—least of all farmers—should be satisfied with agricultural practices that degrade and pollute our waterways. Using less fertilizer would go a long way toward addressing pollution problems, like the one that recently led the Des Moines Water Works, which supplies the city's drinking water, to sue three Iowa farm counties over the nitrates contaminating the water supply. It's safe to say that something is wrong with our agricultural system when neighbors collectively sue those who feed them for poisoning their water. Widespread adoption of conservation agriculture would help solve nitrate, phosphate, and pesticide pollution problems writ small and large, from individual on-farm water wells to the great dead zone in the Gulf of Mexico.

And it is worth considering the inestimable value of soil biodiversity to human health in light of the fact that most modern antibiotics came from soil-dwelling microbes. We are far from knowing all the mircoorganisms that live in native soil communities. Who knows which one may next prove transformational for agriculture or medicine? We need to stop relying on tillage and intensive fertilizer use that bankrupts nature's stores. The accompanying alteration of soil biota reduces diversity and shifts bacterial and fungal community abundance and compositions. Restoring organic matter to soils and adopting practices with less physical and chemical disturbance can counter these problems.

The promise of conservation agriculture to bring life back to the land and support biodiversity both above and belowground should appeal to environmentalists and farmers alike. For like it or not, a large part of nature will be what lives on farms, because we now use more than a third of the world's ice-free land area for growing crops and raising animals.

But just because we *can* restore degraded land rapidly doesn't mean we will. Under conventional practices, an individual farmer often faces a choice between prioritizing short-term profit or conserving soil and its fertility over the long run. Yet as a practical matter, conservation cannot come at the expense of economic viability—any truly sustainable agriculture needs both to align. One of the most promising things about practicing all three elements of conservation agriculture as an agronomic system is that it can save conventional farmers both time and money.

Unlike the fertilizer-intensive Green Revolution practices that developed top-down through government agencies and corporate research, conservation agriculture has largely evolved and spread through bottom-up farmer-led initiatives. Why? A key attraction is the opportunity to improve a farm's bottom line by lowering input costs.

But it's not just farmers who are interested. A number of prominent foundations have adopted soil health as a central theme of their efforts. Chief among them are the Howard G. Buffett Foundation in Illinois, the Noble Foundation in Oklahoma, and the Regenerative Agriculture Foundation in California. And dozens of nonprofit organizations around the world now promote soil health and restoration, including the recently established Soil Health Institute in North Carolina. Even corporate titan Shell Oil is supporting a major test to assess the potential for large-scale carbon sequestration in rangeland soils.

Who are the biggest laggards? From my outsider perch, I'd have to vote for the USDA. While there are influential voices within the agency who are true leaders in the soil health movement, overall the agency isn't vigorously researching or promoting practices to rebuild the fertility of our soil—the foundation for our nation's future.

Hopefully, they're just getting started. In 2012, the Natural Resource

Conservation Service (NRCS) kicked off a national soil health pro-gram to promote knowledge of how practices like no-till, cover crop-ping, and diverse rotations build soil carbon and the microbial activity that underpins higher profits and yields. While this is an influential and welcome development, I got the impression from farmers I talked to that, overall, USDA programs undermine practices that promote and restore soil health, and some indirectly discourage adoption of conservation agriculture practices. There's a lot of inertia to overcome in order to change conventional farming. Yet as I was finishing this book, the White House Office of Science and Technology Policy issued a national call to action, encouraging public and private efforts to pro-tect America's soil. Hopefully the tide is turning, for establishing a national soil health policy should not be a partisan issue.

Still, most academic agricultural research focuses on improving cur-rent methods and practices, rather than investigating the potential of alternative systems. Research on soil health and conservation agriculture systems has been estimated to receive less than 2 percent of agricultural research funding in the United States, and less than 1 percent globally. An analysis of 284 projects in the USDA-supported $294 million Research, Extension, and Economics program budget for 2014 found that projects including cover-cropping for pest control or soil-conditioning received 6 percent, and projects involving complex crop rotations received 3 per-cent. Rotational or regenerative grazing received less than 1 percent, as did research on integrated crop-livestock systems.

The vast majority of funding went to support incremental adjust-ments to conventional methods. Most agronomic research remains focused on testing or developing new products and technical advances in conventional practices. Over and over, farmers emphasized to me how our public research system has been subverted to focus on com-mercial products at the expense of regenerative practices. And yet it is shifts in practices that could prove truly transformative.

Why isn't there greater policy support for encouraging wider adop-tion of conservation agriculture and regenerative farming practices? Asking this, I got various earfuls from farmers. One factor they mentioned is the proverbial revolving door between industry executives and political appointees that run government agencies. And career civil servants know better than to advocate for change ahead of the political winds. I might have gotten a glimpse of this in the look on the face of a highly placed USDA official when I asked what he thought about either making crop insurance available only to farmers practicing conservation agriculture, or simply ditching the program altogether.

Naturally, one of the greatest obstacles is the influence of agribusiness lobbyists who purvey chemical solutions to biological problems. When I asked farmers why there is relatively little federal support for adopting conservation agriculture, most answered along the lines of "follow the money" to the major industries influencing Congress and regulatory agencies. Few were shy about pointing to what they saw as the biggest obstacles to change—turf-defending government programs and commercial interests steering the policy show.

Change will not come easy. Agribusiness is now as much about selling products to farmers as selling what farmers produce. One person I interviewed for this book told a story about one of his graduate students who'd gone back to the family farm one summer. The student worked out that, between higher input costs and lower crop prices, his father and brothers harvested a net profit of just 50 cents an acre the prior year. They would have been better off buying a single packet of pumpkin seeds and hand-planting them, forgoing the cost of inputs. Assuming a single pumpkin survived to harvest on each acre, they could have sold it for six times what they actually ended up earning— and saved themselves all the work of plowing, fertilizing, and harvesting. This story illustrates how the people making most of the money from farming are not the farmers. It's those who sell stuff to farmers who are doing really well under the current system—the companies who sell the inputs on which conventional farming rests.

Today, the margin between losing the farm and staying on the land is pretty tight for most farmers. They can't choose the price they pay for fertilizer, diesel, and all their other inputs, or set the price they get for their corn, wheat, or soybeans. But they can change their practices to reduce their need and expenses for inputs. As I was writing this chapter, I came across a study projecting that 27 percent of row-crop land in Iowa would lose more than $100 an acre in 2015, due to high input costs and falling grain prices. Something is seriously wrong with our agricultural system if hardworking Iowans growing crops on some of the best agricultural soil in the world can't make money farming.

Yet, if conservation agriculture is more profitable, why do the majority of farmers still practice tillage-based high-input farming? There are barriers large and small. In many regions, lack of knowledge about how to adapt conservation agriculture methods to local conditions and crops remains a major obstacle to their adoption. A 2012 FAO review reported that the key to farmers making the shift to conservation agriculture is providing them with local examples of successful implementation at full-scale demonstration farms, coupled with training and technical support for early adopters. Of course, another barrier is the potential economic hit a farmer can take during the transition.

Adopting conservation agriculture involves changes in long-standing cultural practices and a change in mind-set. It represents a new system of farming, the success of which depends more on what farmers do than on their level of input use. Over the past decade, conservation agriculture has been increasingly promoted among smallholder farmers in the tropics. Yet adoption of all three elements remains low; smallholder adoption worldwide remains just a few percent.

Much of the argument around conservation agriculture is not over whether its principles work, but over whether the great number of smallholders around the world will adopt them. Communal grazing, removal of crop residue for cattle fodder, and the use of animal dung for cooking fuel all hamper adoption of conservation agriculture in Africa and Asia. These practices curtail residue retention and thus preclude realizing the potential of the full system. And converting to no-till without adopting the other two elements of cover-cropping and diverse rotations can *reduce* yields.

Lack of access to capital and credit can also limit farmer access to even minimal inputs and seeds for crop diversification and rotation.

Lack of access to direct seeding equipment designed for use with low power equipment can also be a problem. Such farm-level socioeconomic constraints mean that adoption of conservation agriculture practices in Africa often involves adopting just one or two of its elements. Expanded adoption of conservation agriculture would require addressing these and other social challenges.

Barriers in the developed world include misinformation among agricultural authorities trained and steeped in conventional practices and views, as well as farmer resistance to change and prejudice for traditional or conventional methods. In addition, commodity-based subsidies and price supports favor monocultures or simple rotations, and crop insurance rules can discourage farmers from planting cover crops.

While there appears to be a strongly held view among many agronomists that conservation agriculture depends on mechanization, high input levels, and herbicide-resistant crops, the farmers I visited showed me that this is not actually the case. There is a wide range of ways to adapt farming practices to the general goal of improving soil health. Perhaps the label of conservation agriculture has outlived its usefulness in this regard. A more general goal would be to develop agricultural practices that bolster soil health *and* increase yields. This can be done—I've seen it done—on farms large and small, rich and poor, conventional and organic.

So far, however, it's a bottom-up revolution, driven by individual farmers looking for a better margin as they're squeezed between input and commodity pricing beyond their control. I thought farmers would want some top-down help. So I was surprised to hear them argue that making soil restoration a natural consequence of agricultural production would not require government subsidies. Almost all of those I talked to said that if government simply got out of the subsidy business altogether, organic and conservation agriculture farms would outcompete conventional ones. As it is we now subsidize and incentivize practices that degrade soil health.

Crop insurance and government programs for food security started in the Depression Era to protect farmers and ensure a stable food supply. All too often, they now offer farmers an easy out for poor practices, discouraging the kind of innovation that many farmers are particularly good at—problem-solving on their own piece of ground. Farmers themselves consistently volunteered that farming practices would turn on a dime if crop insurance programs incentivized conservation agriculture.

New Zealand already showed that eliminating agricultural subsidies won't lead to catastrophe. Farmers fiercely resisted when the government decided in 1984 to end farm subsidies that accounted for more than a third of gross farm income. The disaster they predicted never struck. Twenty years later, the Federated Farmers of New Zealand published a retrospective report that concluded the move resulted in substantial productivity increases, more diversified land use, and more efficient use of fewer farm inputs, particularly fertilizers, which lowered their production costs. Farmers were no longer chasing subsidies and pursuing maximum yields at any cost. After two decades off subsidies, the farmers themselves concluded that the former subsidies had limited agricultural innovation and productivity.

We in the United States have our subsidies backward. Changing crop insurance programs and subsidies to promote building soil health could better align farmers' short-term interests with society's long-term interests. Why not financially backstop farmers for the first couple of years, during the transition period? At the very least, soil-building practices should get a better payout instead of automatic disqualification, as farmers I talked to repeatedly complained about. Revising commodity support programs to encourage, if not require, cover crops and diverse rotations offers another way to move toward practices that improve soil health. From a societal perspective, it makes sense to restructure agricultural subsidies to reward farmers for improving soil fertility—it makes none to continue subsidizing practices that do the opposite.

While no single metric can capture the complexity of soil health, soil carbon offers an essential, simple way to judge and measure it. Allowing farmers to bank carbon credits through increasing soil organic matter would provide an incentive for investing in soil restoration. Carbon credits could provide an income stream for farmers based on the societal value of carbon sequestration, reduced water pollution, and maintenance of soil fertility and pollinator populations. A 2015 paper by a consortium of European researchers estimated that, on average, a 1 percent loss in soil organic carbon translated to a societal loss of natural capital amounting to about $66 an acre. Rattan Lal estimated the societal value of carbon at $120 per ton. I suspect that if farmers could make that much an acre per year they would flock to adopt soil-building practices.

Big changes in thought in both academia and agencies like the USDA tend to occur as people retire and new people come on board with different training and ideas. It is difficult for authority figures like professors, senior scientists, and technical staff to accept that what they taught for decades is, at best, only part of the story. The recent NRCS push to promote soil health as foundational to both good farming and our national interest should not only continue but receive greatly increased support.

When we were driving across Kansas, Guy Swanson told me that he thinks the key to increasing adoption of no-till is to start with younger farmers who don't already believe they need to plow. It would also help to expand agronomy curricula to include teaching and research on conservation agriculture practices as a viable system of environmentally attractive, economical alternatives to conventional practices. A parallel effort is needed to develop educational literature on conservation agriculture practices to provide to farmers, agricultural universities, and schools around the world.

We also need programs to put young people back on the land—and reward them if they build soil health. The average age of American farmers is about sixty. As the current generation of farmers retires, we need to encourage a new generation to return to the land—and train and equip them to adopt regenerative practices.

When Anne and I were in England, we learned of a novel program to save family farms and establish a new generation on the land. The program paired farmers nearing retirement who had no family member interested in taking over the farm with young people who wanted to be farmers but lacked the means to buy a farm. The older farmers would teach the younger ones, and the younger farmers would build equity through their labor and eventually buy out their retiring mentors.

A North American version of such a program might be thought of as a twenty-first-century Homestead Act. Another possibility would be to set up a public land bank for foreclosed farms and agricultural land to facilitate young people buying a farm with a long-term investment of labor. The deal might be that if this new kind of homesteader-farmer improves their soil enough over twenty years, the farm becomes his or hers. Such a program could be used to help preserve enough fertile farmland near cities to feed urban populations well into the future, something in everyone's interest.

Another idea would be to update what worked well for centuies and reintegrate crop and livestock production on farms small and large. In particular, this offers a major opportunity to turn manure from hazardous feedlot waste into a valuable tool for building soil health. In this case, concentration really is the problem, and dilution really is the solution. We need to work manure back into the soil on *a lot* of farms instead of dumping it all in centralized locations to form toxic lagoons.

Of course, spreading composted manure back on agricultural fields should motivate rethinking what we feed cows. Most immediate and critical is ending the use of antibiotics for growth promotion and other unintended purposes. And as far as feeding the world goes, it's obvious that grazing cattle on crop stubble is better than raising them on crops we humans can eat.

The path to promoting the return of livestock to smaller diversified farms lies in rebuilding decentralized infrastructure like small-scale slaughter facilties to make meat production and processing easier for small farms. Streamlined permitting and regulatory processes for small farms and packing facilities would also help. Gabe Brown pointed out to me that big salmonella outbreaks tend to come from large meat processors whose constantly running, large-scale operations are challenging to clean. In this arena, I can see sound policy and public health reasons for favoring the little guy over big operators. Of course, the big packers have more money and lobbyists than do small farmers.

Finally, another obstacle is that at present there is no way for consumer demand to support soil health, other than by consumers buying organic. But this doesn't always match up well. After all, organic practices do not necessarily improve or maintain soil health—it depends on the organic farmer's practices, especially tillage. Informed consumers are one of the best and fastest ways to move the commercial dial in market economies. If consumers realized that their health is intimately connected to the quality and fertility of soil—for better or worse—this could help move conventional farming toward more sustainable, organic-ish practices. So how can we brand soil health to provide consumers the information necessary to make the well-informed choices they deserve and desire? I suspect that the best approach would be some kind of "soil safe" certification from a national association of regenerative, soil-building farmers.

Achieving the full potential of a soil health revolution will take every tool in our agrotechnology toolkit—and a lot of open-minded experiments to adapt and amend the general principles of conservation agriculture to specific farms, soils, and crops. Farmers typically say they want to leave the soil better than they found it, but they don't always know how to go about doing it. What makes for a good rotation in Missouri will not necessarily work well for Pennsylvania, or eastern Washington. Still, however one looks at it, corn-soy is not a complex rotation. But if that's what you've been growing—what your crop advisors have recommended you grow year after year—what should you grow instead? A common element in the successful adoption of conservation agriculture in regions I visited was the value of regional demonstration farms in promoting farmer adoption.

Farms like Dakota Lakes and Kofi Boa's No-Till Center offer models for how to make agricultural research more relevant to farmers' needs. Demonstration farms can show farmers how to adopt new systems without betting their own farms. What is the best way to establish demonstration farms focused on soil health? In addition to the farmer-owned co-op model of Dakota Lakes, another is to use the national network of conservation districts. Almost every county in the United States has one, established under state law and known by different names in different states. Menoken Farm in Burleigh County, North Dakota, shows how well this can work. A private-sector model offers a third—like Kofi Boa's No-Till Center and the Rodale Institute—and publicly supported farms are potentially a fourth model. What is universal is the need to establish such farms. A global network of demonstration farms dedicated to pursuing regional recipes for regenerative agriculture is one of the best investments humanity could make in our own future.

On my travels I heard a lot about how most current agricultural research is largely irrelevant to practicing or adopting conservation agriculture. A common complaint is that academic researchers shun applied research or won't try to adapt methods to new areas or settings. Another called out small-scale research plots that farmers find unconvincing, and the organization of agronomic science into disciplinary silos that discourage the type of cross-disciplinary, system-level thinking, insights, and research that underpin conservation agriculture. We need more agronomists and soil scientists working together with soil biologists, entomologists, and others. These are all things that demonstration farms could facilitate and do well.

When Soviet premier Nikita Khrushchev flew nonstop from Moscow to Washington on September 15, 1959, Americans were stunned. Our planes couldn't do that! At dinner that evening Khrushchev went a step further and gave President Eisenhower a replica of the Soviet Lunik 2 probe. To much fanfare, it had landed on the moon the day before. How could we top that? Three years later, President Kennedy declared that America would put a man on the moon by the end of the decade—and bring him back safely.

As I write this, we're approaching the fiftieth anniversary of Apollo 11, and I see some parallels. We need to bring the same intensity and focus to bear on transforming conventional agriculture.

How can we do this? Adopt policies that promote soil-building practices. We need a soil health moonshot, an era of public investment in research and incentives to change the business of agriculture and secure the living foundation for our future on this planet. Obviously, the private sector would be involved in such an endeavor, but private corporations are unlikely to spearhead and support research into practices that use fewer of their products. Yet soil health should serve as the new lens through which to evaluate agricultural science, practices, and technologies. We need to train a new generation of systems thinkers and support research on practices as well as products—with an emphasis on strategic partnerships with farmers, the people who actually grow our food, fodder, and fiber.

Our understanding of the world—and of soil—has changed dramatically over time, and could do so again. Since the dawn of agriculture, people have seen soil as something to be worked, an arena in which human labor could harness and tame nature. Societies around the world cast the great mystery of soil fertility as a gift of the gods, with harvests subject to change on a whim or a prayer to the Greek Demeter, the Roman Ceres, or the Hindu Lakshmi. And for many today, as then, the fertility of the soil remains the key to a livelihood.

With the advent of the Renaissance, soil offered a decipherable mystery that could be understood through the application of reason. As natural philosophers began to contemplate its secrets, Leonardo da Vinci famously wrote, "We know more about the stars overhead than the soil underfoot." His words still ring true today, more than five hundred years later.

When early agronomists began investigating soil fertility and husbandry, crop rotations and animal manure became central to improving land and building fertile soil. But these ideas lost their luster in the nineteenth century with the discovery of the power of chemical fertilizers to boost crop yields on degraded land. The near-miraculous effects of these new chemical supplements gave rise to the view of soil as little more than a physical receptacle for agrochemicals, a reservoir or gas tank to be topped off as needed. The mechanization that then reshaped agriculture led people to increasingly view soil as the least expensive—and least valuable—input in industrial crop production. As we've seen, this perspective has done grievous harm to the world's soil—civilization's foundation.

Views of soil fertility are changing once again as we begin to accept that it depends on soil biology as much as soil chemistry and physics. And while we still have much to learn, recent discoveries reveal that soil ecology holds the key to nutrient availability and cycling, and to maintaining soil fertility. Now that we know the critical role of soil life, we can see the necessity of viewing soils rich in organic matter as an essential part of nature's grand cycle of growth and decay.

Perhaps we can turn to history for some inspiration. Our founding farmers Jefferson and Washington followed two out of the three principles of conservation agriculture, but relied on the plow for planting and weed control. They almost got it right. Today we have no-till planters and other means of weed control, the third leg we need to stabilize the agricultural stool.

The essence of this new revolution-in-the-making isn't complicated. It boils down to two words—soil health. On the ground this means prioritizing agricultural practices that build soil organic matter. Yet farmers don't need to go organic to lead the charge or play a supporting role. Agrochemicals can be useful tools—when used wisely. But relying on them to substitute for healthy fertile soil really does live up to the "more on" moniker. We need to embrace a new philosophy that evaluates farming practices by assessing whether they build or mine soil fertility.

After centuries of degrading the soil upon which our continued livelihood depends, we need to reinvest in our most fundamental resource if our global civilization is to avoid the fate of prior regional ones. At a basic level this sounds pretty simple. The problem is that we have to do it on national and global scales. And for this we need a new system of farming, an agricultural system that yields not just bountiful harvests but improved soil health.

Part of the reason the Green Revolution worked so well was that we'd already seriously degraded soil fertility through killing off soil life with mechanical disturbance and chemical inputs. Substituting fertilizers and pesticides for soil life compensated for this lost fertility. Now, we need a new philosophy of conventional farming, a fundamental rethinking of the basic principles behind how we do it. Brain transplants, as Dwayne Beck more colorfully puts it. We need changes in practices as much, if not more, than we need new technologies. Of course, technology and agrochemicals can help, but they provide tools that can be employed in a good or bad system. And I know we can farm smarter. Farmers I visited already do.

Their revolutionary approach might best be captured as: ditch the plow, cover up, and grow diversity. These regenerative agricultural practices don't require cutting-edge technology or waiting to invent something new. They are ready to go and scalable to small- and large-sized farms using existing technology in both the developed and developing worlds. As we've seen, innovative farmers already following these principles demonstrate that they can work better for both the land and those who work it.

The convergence of new science, declining resource availability, and a rising population calls for creative solutions. Fortunately, conservation agriculture offers an already demonstrated way to increase crop yields that is not yet widely used. Its transformative potential lies in adopting all three of its underlying principles and recognizing the need to develop practices well suited to different soils, climates, crops, and even individual farms. And while soil health is no silver bullet, it's becoming less of a secret weapon as the farmers that have abandoned conventional agriculture for regenerative practices see that this allows them to grow more with less.

Yes, we really can change the world and write a new ending to an ancient story. For fertile soil can be lost through—or result from—how we farm. I find it fitting that *humus* and *human* share the same Latin root, as restoring healthy soil to the world's agricultural lands is one of the best investments we can make in humanity's future. And so as we grapple with the daunting problems of how to feed the world, cool the planet, and stem losses in the natural world, let's not lose sight of a simple truth. Sometimes answers we seek are closer than we might think—right beneath our feet.

ACKNOWLEDGMENTS

Foremost I'd like to thank the farmers and scientists who allowed me to visit their farms and gave generously of their time and support so that I and others might learn from their experience—Ray Archuleta, Mike and Anne Arnoldy, Herbert Bartz, Dwayne Beck, Alejandro Biamonte, Kofi Boa, David Brandt, Gabe and Paul Brown, Sally Brown, Howard G. Buffett, Mike Cronin, Neil and Barbara Dennis, Rolf Derpsch, Dan Eberhardt, Felicia Echeverría, Dan Forgey, Jay Fuhrer, Ralph and Betty Holzwarth, Elaine Ingham, Wes Jackson, Kent Kinkler, Peter Kring, Rattan Lal, Alex and Leilani Lamberts, Jonathan Lundgren, Joel McClure, Javier Meza, Jeff Moyer, Kristine Nichols, Emmanuel Omondi, Don Reicosky, Marv Schumacher, and Guy Swanson. Without their patience, hospitality, and knowledge I would never have been able to write this book.

Once again, my wife Anne not only put up with me taking on another book, but helped me sharpen up—and shorten—chapter drafts. As ever, her insights, advice, and suggestions seriously improved the final product.

Every writer knows that it takes a team to produce a book, and I'm fortunate to have a great one. My agent, Elizabeth Wales, helped frame the idea for the book and set me on the right path. Ingrid Emerick and Anna Katz of Girl Friday Productions helped focus, shape, and trim the manuscript as it took shape. Maria Guarnaschelli, my editor at W. W. Norton, championed it from conception to completion and Nathaniel Dennett, her assistant, shepherded the book through the publishing process. I also thank Fred Wiemer for his excellent copy editing.

Several other folks deserve special thanks. Sally Brown helped put this whole thing in motion when she recommended me for the "Dig It" symposium where I met Rattan Lal. Art Donnelly introduced me to biochar-making and was an exemplary guide and traveling companion in Costa Rica. Birgit Lenderink hospitably put up two bedraggled travelers in San José. Heidi Fitzgerald and Klinton Caillier provided an awesome tour of the City of Tacoma's wastewater treatment facility and explained the municipal alchemy they perform there everyday. At the meeting in Malmö, Katarina Hedlund and Mary Scholes guided me to some key sources that were most helpful. And I greatly appreciate the help of Kwasi Boa and Kyei Baffour in getting around in, and getting to know, Ghana. Hearing economist Jeffrey Sachs speak about sustainable development greatly influenced my thinking on the subject. And lectures by Greg Kester, Kuldip Kumar, Sally Brown, Roberta King, and Ian Pepper at the Northwest Biosolids Management Association meeting were especially informative.

While I remain deeply appreciative of all those who helped me pull this book together, and acknowledge my debts to them for good ideas I've assimilated, any errors that have crept into the final product remain mine alone.

For readers interested in digging deeper, I have listed sources in the following pages that I drew upon and consulted in writing the chapters. These include studies referred to in the book and additional research that illustrates key points made or that supports the broader generality of the experiences of farmers I visited.

NOTES

1. Soil quality refers to the inanimate portion of the soil, whereas soil health also incorporates the state of soil life.
2. National Research Council, Committee on the Role of Alternative Farming Methods in Modern Production Agriculture. 1989. *Alternative Agriculture*. Washington, D.C.: National Academy Press, p. 9.
3. "Pesticide" is a general term for biocides that include herbicides, insecticides, fungicides, and a host of other substances meant for destroying pests.
4. National Research Council, Committee on Genetically Engineered Crops. 2016. *Genetically Engineered Crops: Experiences and Prospects*. Washington, D.C.: National Academy Press, p. 65.
5. Bt corn is genetically modified to produce Bt delta endotoxin, a protein borrowed from the soil bacterium *Bacillus thuringiensis* and which kills the larvae of the European corn borer and corn earworms. The toxin is more selective than broad-spectrum insecticides and is not considered harmful to insects in other orders.
6. Howard, A. 1940. *An Agricultural Testament*. London: Oxford University Press, p. 168.
7. Plato. *Timaeus and Critias*. London: Penguin Books, 1977. *Critias*, 111, p. 134.
8. These would be those by Cato (234–149 B.C.), Varro (116–27 B.C.), Columella (A.D. 4–ca. 70), and Pliny the Elder (A.D. 23–79).
9. Washington, G., 1892, *The Writings of George Washington*. Edited by W. C. Ford. Vol. 13. New York: Putnam, pp. 328–329.
10. A note here about terms. There are inconsistencies in the use of the term "conservation agriculture" that cloud academic debates over the effectiveness of conservation agriculture as a system of practices. A number of studies, for example, treat no-till farming as a form of conservation agriculture rather than as a single component of a three-component system. I will use the term "conservation agriculture" through the rest of this book to refer to agricultural practices that follow all three principles: (1) minimal soil disturbance,

(2) integration of cover crops (including legumes), and (3) diverse rotations. However, the practices constituting minimal disturbance and diversity in a rotation are not crisply defined, and this has contributed to confusion in broad analyses of conservation agriculture's effectiveness and generality. The related term "regenerative agriculture" refers to farming practices that build soil and restore fertility as a consequence of crop and/or livestock production.

11. In 1994, Congress changed the agency's name to the Natural Resources Conservation Service.

12. Faulkner, E. 1943. *Plowman's Folly.* Norman: University of Oklahoma Press, p. 128.

13. Lal, R., D. C. Reicosky, and J. D. Hanson. 2007. Evolution of the plow over 10,000 years and the rationale for no-till farming. *Soil & Tillage Research* 93: p. 6.

14. Derpsch, R., et al. 2014. Why do we need to standardize no-tillage research? *Soil & Tillage Research* 137: p. 20.

15. The ALS enzyme (acetolactate synthase) is essential for plant growth. ALS inhibitors act by binding to the ALS enzyme and inactivating it.

16. Green, J. M., and M. D. K. Owen. 2011. Herbicide-resistant crops: Utilities and limitations for herbicide-resistant weed management. *Journal of Agricultural and Food Chemistry* 59: p. 5827.

17. If, like me, you are not familiar with teff (*Eragrostis tef*), it is an annual grass native to Ethiopia that is high in dietary fiber and iron. It is one of humanity's earliest domesticated plants and is attractive to those on low-gluten diets. It also provides high-quality forage for livestock.

18. For those interested in numbers, the erosion rate from the no-till plot for that year was 77 kg/ha/yr, whereas that from the slash-and-burn plot was 1787 kg/ha/yr.

19. Leighty, C. E. 1938. Crop rotation. In *Soils & Men, Yearbook of Agriculture 1938.* 75th Congress, 2nd Session, House Document No. 398, United States Department of Agriculture. Washington, D.C.: Government Printing Office, p. 417.

20. Ibid., p. 411.

21. Ammonia and ammonium nitrate can be used as either fertilizer or to produce high explosives.

22. While the early design was developed by the USDA Agricultural Research Service in Alabama, the most popular current version is based on the Rodale design.

23. Ponisio, L. C., et al. 2015, Diversification practices reduce organic to conventional yield gap. *Proceedings of the Royal Society B* 282: 20141396, doi:10.1098/rspb.2014.1396, p. 4.

24. Manson, M. 1899. Observations on the denudation of vegetation—a suggested remedy for California. *Sierra Club Bulletin* 2: p. 300.

25. *Terra preta* is Portuguese for "black earth."

26. *Bokashi* is Japanese for "fermented organic matter."

27. Coffee pulp is the part of the coffee fruit that surrounds the seeds, or coffee beans. The pulp makes up around 40 percent of the weight of the coffee fruit.

28. The gene for producing the latter's toxin is the basis for genetically engineered Bt crops.

29. Coffee bean chaff is the dried skin of the coffee bean that comes off during the roasting process. Coffee roasters generally consider it a nuisance, but it also makes good mulch, compost, and chicken coop litter.

30. Albrecht, W. A. 1938. Loss of soil organic matter and its restoration. In *Soils & Men, Yearbook of Agriculture 1938*. 75th Congress, 2nd Session, House Document No. 398, United States Department of Agriculture, Washington, D.C.: Government Printing Office, p. 355.

31. Liebig, J. V. 1863. *The Natural Laws of Husbandry*. New York: Appleton, p. 181.

32. Ibid., p. 182.

SOURCES

1 FERTILE RUINS

Alavanja, M. C. R., M. K. Ross, and M. R. Bonner, 2013. Increased cancer burden among pesticide applicators and others due to pesticide exposure. *CA: A Cancer Journal for Clinicians* 63:120–142.

Alexander, E. B. 1988. Rates of soil formation: Implications for soil-loss tolerance. *Soil Science* 145:37–45.

Beard, J. D., et al. 2014. Pesticide exposure and depression among male private pesticide applicators in the Agricultural Health Study. *Environmental Health Perspectives* 122:984–991.

Brink, R. A., J. W. Densmore, and G. A. Hill, 1977. Soil deterioration and the growing world demand for food. *Science* 197:625–630.

Brown, L. R. 1981. World population growth, soil erosion, and food security. *Science* 214:995–1002.

Hooke, R. LeB. 2000. On the history of humans as geomorphic agents. *Geology* 28:43–46.

Intergovernmental Technical Panel on Soils, L. Montanarella, chair. 2015. *Status of the World's Soil Resources: Technical Summary.* Food and Agriculture Organization of the United Nations (FAO), Rome, 79 pp.

International Assessment of Agricultural Knowledge, Science and Technology for Development (IAASTD). Edited by B.D. McIntyre et al. 2009. *Agriculture at a Crossroads.* Washington, D.C.: Island Press, 590 pp.

Lal, R., and B. A. Stewart. 1992. Need for land restoration. In *Soil Restoration.* Edited by R. Lal and B. A. Stewart. *Advances in Soil Science,* vol. 17, pp. 1–11.

Larson, W. E., F. J. Pierce, and R. H. Dowdy. 1983. The threat of soil erosion to long-term crop production. *Science* 219:458–465.

Montgomery, D. R. 2007. Soil erosion and agricultural sustainability, *Proceedings of the National Academy of Sciences* 104:13,268–13,272.

Pimentel, D., et al. 1987. World agriculture and soil erosion. *BioScience* 37:277–283.

Pimentel, D., et al. 1995. Environmental and economic costs of soil erosion and conservation benefits. *Science* 267:1117–1123.

Wakatsuki, T., and A. Rasyidin. 1992. Rates of weathering and soil formation. *Geoderma* 52:51–63.

Wilkinson, B. H., and B. J. McElroy. 2006. The impact of humans on continental erosion and sedimentation. *Geological Society of America Bulletin* 119:140–156.

2 MYTHS OF MODERN AGRICULTURE

Appenzeller, T. 2004. The end of cheap oil. *National Geographic* 205, no. 6:80–109.

Battisti, D. S., and R. L. Naylor. 2009. Historical warnings of future food insecurity with unprecedented seasonal heat. *Science* 323:240–244.

Benbrook, C. M. 2012. Impacts of genetically engineered crops on pesticide use in the U.S.—the first sixteen years. *Environmental Sciences Europe* 24:24, doi:10.1186/2190-4715-24-24.

Erdkamp, P. 2002. 'A starving mob has no respect': Urban markets and food riots in the Roman world, 100B.C.—400A.D. In L. de Blois and J. Rich, eds. *The Transformation of Economic Life Under the Roman Empire*. Amsterdam: Geiben, pp. 93–115.

Fedoroff, N. V., et al. 2010. Radically rethinking agriculture for the 21st century. *Science* 327:833–834.

Foley, J. A., et al. 2005. Global consequences of land use. *Science* 309:570–574.

Foley, J. A., et al. 2011. Solutions for a cultivated planet. *Nature* 478:337–342.

Godfray, H. C. J., et al. 2010. Food security: The challenge of feeding 9 billion people. *Science* 327:812–818.

Lipinski, B., et al. 2013. *Reducing Food Loss and Waste*, World Resources Institute & United Nations Environment Program, 39 pp.

Lobell, D. B., and C. Tebaldi. 2014. Getting caught with our plants down: The risks of a global crop yield slowdown from climate trends in the next two decades. *Environmental Research Letters* 9:074003, doi:10.1088/1748-9326/9/7/074003.

McGlade, C., and P. Ekins. 2015. The geographical distribution of fossil fuels unused when limiting global warming to 2°C. *Nature* 517:187–190.

National Research Council, Committee on Genetically Engineered Crops. 2016. *Genetically Engineered Crops: Experiences and Prospects*. Washington, D.C.: National Academy Press, 388 pp.

National Research Council, Committee on the Role of Alternative Farming Methods in Modern Production Agriculture. 1989. *Alternative Agriculture*. Washington, D.C.: National Academy Press, 448 pp.

Pimentel, D., et al. 1976. Land degradation: Effects on food and energy resources. *Science* 194:149–155.

Ray, D. K. et al. 2012. Recent patterns of crop yield growth and stagnation. *Nature Communications* 3:1293, doi:10.1038/ncomms2296.

Ruttan, V. W. 1999. The transition to agricultural sustainability. *Proceedings of the National Academy of Sciences* 96: 5960–5967.

Scholes, M. C., and R. J. Scholes. 2013. Dust unto dust. *Science* 342: 565–566.

Tilman, D., et al. 2001. Forecasting agriculturally driven global environmental change. *Science* 292: 281–284.

Tilman, D., et al. 2002. Agricultural sustainability and intensive production practices. *Nature* 418:671–677.

3 ROOTS OF THE UNDERGROUND ECONOMY

Altieri, M. A., and C. I. Nicholls. 2003. Soil fertility management and insect pests: Harmonizing soil and plant health in agroecosystems. *Soil & Tillage Research* 72: 203–211.

Howard, A. 1940. *An Agricultural Testament*. London: Oxford University Press.

Humphreys, C. P., et al. 2010. Mutualistic mycorrihiza-like symbiosis in the most ancient group of land plants. *Nature Communications* 1:103, doi:10/1038/ncomms1105.

Khan, S. A., et al. 2007. The myth of nitrogen fertilization for soil carbon sequestration. *Journal of Environmental Quality* 36:1821–1832.

Kimpinski, J., and A. V. Sturz. 2003. Managing crop root zone ecosystems for prevention of harmful and encouragement of beneficial nematodes. *Soil & Tillage Research* 72:213–221.

Kremer, R. J., and J. Li. 2003. Developing weed-suppressive soils through improved soil quality management. *Soil & Tillage Research* 72:193–202.

Montgomery, D. R., and A. Biklé. 2016. *The Hidden Half of Nature: The Microbial Roots of Life and Health*. New York: W. W. Norton, 309 pp.

Mulvaney, R. L., S. A. Khan, and T. R. Ellsworth. 2009. Synthetic nitrogen fertilizers deplete soil nitrogen: A global dilemma for sustainable cereal production. *Journal of Environmental Quality* 38:2295–2314.

Schomberg, H. H., P. B. Ford, and W. L. Hargrove. 1994. Influence of crop residues on nutrient cycling and soil chemical properties. In *Managing Agricultural Residues*. Edited by P. W. Unger. Boca Raton, FL: Lewis Publishers, pp. 99–121.

Tiessen, H., E. Cuevas, and P. Chacon. 1994. The role of soil organic matter in sustaining soil fertility. *Nature* 371:783–785.

Wallace, A. 1994. Soil acidification from use of too much fertilizer. *Communications in Soil Science and Plant Analysis* 25:87–92.

4 THE OLDEST PROBLEM

Betts, E. M., ed. 1953. *Thomas Jefferson's Farm Book*. Princeton: Princeton University Press, 552 pp.

Craven, A. O. 1925. *Soil Exhaustion as a Factor in the Agricultural History of Virginia and Maryland, 1606–1860*. University of Illinois Studies in the Social Sciences 13, no. 1. Urbana: University of Illinois.

Glenn, L. C. 1911. *Denudation and Erosion in the Southern Appalachian Region and the Monongahela Basin*, U. S. Geological Survey Professional Paper 72. Washington, D.C.: Government Printing Office.

Hughes, J. D., and J. V. Thirgood. 1982. Deforestation, erosion, and forest management in ancient Greece and Rome. *Journal of Forest History* 26:60–75.

Judson, S. 1963. Erosion and deposition of Italian stream valleys during historic time. *Science* 140:898–899.

Judson, S. 1968. Erosion rates near Rome, Italy. *Science* 160:1444–1446.

Lowdermilk, W. C. 1953. *Conquest of the Land Through 7,000 Years*. Agricultural Information Bulletin No. 99, U.S. Department of Agriculture, Soil Conservation Service. Washington D.C.: Government Printing Office, 30 pp.

Marsh, G. P. 1864. *Man and Nature; or, Physical Geography as Modified by Human Action*. New York: Scribner, 560 pp.

Meade, R. H. 1982. Sources, sinks, and storage of river sediment in the Atlantic drainage of the United States. *Journal of Geology* 90:235–252.

Montgomery, D. R. 2007. *Dirt: The Erosion of Civilizations*. Berkeley: University of California Press, 285 pp.

Plato. *Timaeus and Critias*. London: Penguin Books, 1977.

Runnels, C. N. 1995. Environmental degradation in Ancient Greece. *Scientific American* 272:96–99.

Runnels, C. 2000. Anthropogenic soil erosion in prehistoric Greece: The contribution of regional surveys to the archaeology of environmental disruptions and human response. In *Environmental Disaster and the Archaeology of Human Response*. Edited by R. M. Reycraft and G. Bawen. Maxwell Museum of Anthropology, Anthropological Papers 7. Albuquerque: University of New Mexico, pp. 11–20.

Reusser, L., P. Bierman, and D. Rood, 2015. Quantifying human impacts on rates of erosion and sediment transport at a landscape scale. *Geology* 43:171–174.

Stuiver, M. 1978. Atmospheric carbon dioxide and carbon reservoir changes: Reduction in terrestrial carbon reservoirs since 1850 has resulted in atmospheric carbon dioxide increases. *Science* 199:253–58.

Trimble, S. W. 1977. The fallacy of stream equilibrium in contemporary denudation studies. *American Journal of Science* 277:876–887.

Van Andel, T. H., and C. Runnels. 1987. *Beyond the Acropolis: A Rural Greek Past*. Stanford, CA: Stanford University Press, 236 pp.

Van Andel, T. H., C. N. Runnels, and K. O. Pope. 1986. Five thousand years of land use and abuse in the Southern Argolid, Greece. *Hesperia* 55:103–128.

Van Andel, T. H., E. Zongger, and A. Demitrack. 1990. Land use and soil erosion in prehistoric and historical Greece. *Journal of Field Archaeology* 17:379–396.

Washington, G., 1892, *The Writings of George Washington*. Edited by W. C. Ford. Vol. 13. New York: Putnam.

5 DITCHING THE PLOW

Baumhardt, R. L., B. A. Stewart, and U. M. Sainju. 2015. North American soil degradation: Processes, practices, and mitigating strategies. *Sustainability* 7:2936–2960.

Bennett, H. H., and W. R. Chapline. 1928. *Soil Erosion, A National Menace*. U.S. Department of Agriculture, Bureau of Chemistry and Soils and Forest Service, Circular 3. Washington, D.C.: Government Printing Office.

Derpsch, R., et al. 2010. Current status of adoption of no-till farming in the world and some of its main benefits. *International Journal of Agricultural and Biological Engineering* 3:1–26.

Derpsch, R., et al. 2014. Why do we need to standardize no-tillage research? *Soil & Tillage Research* 137:16–22.

FAO. 1996 and 2000, *The Production Yearbook*. Rome.

Farooq, M., and K. H. M. Siddique. 2015. Conservation agriculture: Concepts, brief history, and impacts on agricultural systems. In M. Farooq, and K. H. M. Siddique, eds. *Conservation Agriculture*, Springer International Publishing, pp. 3–17.

Faulkner, E. 1943. *Plowman's Folly*. Norman: University of Oklahoma Press, 155 pp.

Fisher, R. A., F. Santiveri, and I. R. Vidal. 2002. Crop rotation, tillage, and crop residue management for wheat and maize in the sub-humid tropical highlands, II. Maize and system performance. *Field Crops Research* 79:123–137.

Fukuoka, M. 1978. *The One-straw Revolution*. Emmaus, PA: Rodale Press, 181 pp.

Glover, J. D., et al. 2010. Increased food and ecosystem security via perennial grains. *Science* 328:1638–1639.

Greenland, D. J. 1975. Bringing the Green Revolution to the shifting cultivator. *Science* 190:841–844.

Greenland, D. J., and R. Lal, eds. 1977. *Soil Conservation and Management in the Humid Tropics*. Chichester, UK: John Wiley & Sons, 283 pp.

Hobbs, P. R., K. Sayre, and R. Gupta. 2008. The role of conservation agriculture in sustainable agriculture. *Philosophical Transactions of the Royal Society B* 363:543–555.

Jack, W. T. 1946. *The Furrow and Us*. Philadelphia: Dorrance and Co., 158 pp.

Jackson, W. 1980. *New Roots for Agriculture*. Lincoln: University of Nebraska Press, 151 pp.

Jackson, W. 2002. Natural systems agriculture: A truly radical alternative. *Agriculture, Ecosystems & Environment* 88:111–117.

Jat, R. A., K. L. Sahrawat, and A. H. Kassam, eds. 2014. *Conservation Agriculture: Global Prospects and Challenges*. Wallingford (UK) and Boston: CAB International, 393 pp.

Junior, R. C., A. G. de Araújo, and R. F. Llanillo. 2012. *No-Till Agriculture in Southern Brazil: Factors that Facilitated the Evolution of the System and the Development of the Mechanization of Conservation Farming*. United Nations Food and Agriculture Organization and Agricultural Research Institute of Paraná State, 77 pp.

Kassam, A., R. Derpsch, and T. Friedrich. 2014. Global achievements in soil and water conservation: The case of Conservation Agriculture. *International Soil and Water Conservation Research* 2:5–13.

Kassam, A., and T. Friedrich. 2012. An ecologically sustainable approach to agricultural production intensification: Global perspectives and developments. *Field Actions Science Reports*, Special Issue 6: http://factsreports.revues.org/1382.

Kassam, A., et al. 2015. Overview of the worldwide spread of conservation agriculture, *Field Actions Science Reports* 8: http://factsreports.revues.org/3966.

Lal, R. 1976. No tillage effects on soil properties under different crops in western Nigeria. *Soil Science Society of America Proceedings* 7:762–768.

Lal, R. 1976. Soil erosion on alfisols in Western Nigeria, I: Effects of slope, crop rotation, and residue management. *Geoderma* 16:363–375.

Lal, R. 1976. Soil erosion on alfisols in Western Nigeria, II: Effect of mulch rate. *Geoderma* 16:377–387.

Lal, R. 2004. Historical development of no-till farming. In R. Lal, P. R. Hobbs, N. Uphoff, and D. O. Hansen, eds, *Sustainable Agriculture and the International Rice-Wheat System*. New York and Basel: Marcel Dekker, pp. 55–82.

Lal, R. 2009. The plow and agricultural sustainability. *Journal of Sustainable Agriculture* 33:66–84.

Lal, R., D. C. Reicosky, and J. D. Hanson. 2007. Evolution of the plow over 10,000 years and the rationale for no-till farming. *Soil & Tillage Research* 93:1–12.

Lal, R., P. A. Sanchez, and R. W. Cummings, Jr., eds. 1986. *Land Clearing and Development in the Tropics*. Rotterdam & Boston: A. A. Balkema, 450 pp.

Li, H. W., et al. 2007. Effects of 15 years of conservation tillage on soil structure and productivity of wheat cultivation in northern China. *Australian Journal of Soil Research* 45:344–350.

Pittelkow, C. M., et al. 2014. Productivity limits and potentials of the principles of conservation agriculture. *Nature* 517:365–368.

Reicosky, D., and C. Crovetto. 2014. No-till systems on the Chequen Farm in Chile: A success story in bringing practice and science together. *International Soil and Water Conservation Research* 2:66–77.

Reicosky, D. C., and M. J. Lindstrom. 1993. Fall tillage method: Effect on short-term carbon dioxide flux from soil. *Agronomy Journal* 85:1237–1243.

Schlessinger, W. H. 1985. Changes in soil carbon storage and associated properties with disturbance and recovery. In J. R. Trabalha and D. E. Reichle, eds. *The Changing Carbon Cycle: A Global Analysis*. New York: Springer-Verlag, pp. 194–220.

Wood, A. 1950. *The Groundnut Affair*. London: Bodley Head, 264 pp.

Yoder, D. C., et al. 2005. No-till transplanting of vegetable and tobacco to reduce erosion and nutrient surface runoff. *Journal of Soil and Water Conservation* 60:68–72.

6 GREEN MANURE

Anderson, R. L. 2000. A cultural systems approach eliminates the need for herbicides in semiarid proso millet. *Weed Technology* 14:602–607.

Anderson, R. L. 2003. An ecological approach to strengthen weed management in the semiarid Great Plains. *Advances in Agronomy* 80:33–62.

Anderson, R. L. 2004. Impact of subsurface tillage on weed dynamics in the Central Great Plains. *Weed Technology* 18:186–192.

Anderson, R. L. 2008. Diversity and no-till: Keys for pest management in the U.S. Great Plains. *Weed Science* 56:141–145.

Anderson, R. L., and D. L. Beck. 2007. Characterizing weed communities among various rotations in Central South Dakota. *Weed Technology* 21:76–79.

Beck, D. L., and R. Doerr. 1992. *No-till Guidelines for the Arid and Semi-Arid Prairies.* South Dakota State University, Agriculture Experiment Station, Bulletin 712, 30 pp.

Douglas, M. R., J. R. Rohr, and J. F. Tooker. 2015. Neonicotinoid insecticide travels through a soil food chain, disrupting biological control on non-target pests and decreasing soya bean yield. *Journal of Applied Ecology* 52:250–260.

Drinkwater, L. E., P. Wagoner, and M. Sarrantonio. 1998. Legume-based cropping systems have reduced carbon and nitrogen losses. *Nature* 396:262–265.

Furlan, L., and D. Kreutzweiser. 2015. Alternatives to neonicotinoid insecticides for pest control: Case studies in agriculture and forestry. *Environmental and Pollution Research* 22:135–147.

Gibbons, D., C. Morrissey, and P. Mineau. 2015. A review of the direct and indirect effects of neonicotinoids and fipronil on vertebrate wildlife. *Environmental and Pollution Research* 22:103–118.

Goulson, D., et al. 2015. Bee declines driven by combined stress from parasites, pesticides, and lack of flowers. *Science* 347:1435.

Hansen, N. C., et al. 2012. Research achievements and adoption of no-till, dryland cropping in the semi-arid U.S. Great Plains. *Field Crops Research* 132:196–203.

Hartwig, N. L., and H. U. Ammon. 2002. Cover crops and living mulches. *Weed Science* 50:688–699.

Hudson, B. D. 1994. Soil organic matter and available water capacity. *Journal of Soil and Water Conservation* 49:189–194.

Liebig, et al. 2002. Crop sequence and nitrogen fertilization effects on soil properties in western Corn Belt. *Soil Science Society of America Journal* 66:596–601.

Pisa, L. W., et al. 2015. Effects of neonicotinoids and fipronil on non-target invertebrates. *Environmental and Pollution Research* 22:68–102.

Razon, L. F. 2014. Life cycle analysis of an alternative to the Haber-Bosch process: Non-renewable energy usage and global warming potential of liquid ammonia from cyanobacteria. *Environmental Progress & Sustainable Energy* 33:618–624.

Tallaksen, J., et al. 2015. Nitrogen fertilizers manufactured using wind power: Greenhouse gas and energy balance of community-scale ammonia production. *Journal of Cleaner Production* 107:626–635.

Van der Sluijs, J. P., et al. 2015. Conclusions of the Worldwide Integrated Assessment on the risks of neonicotinoids and fipronil to biodiversity and ecosystem functioning. *Environmental Science and Pollution Research* 22:148–154.

Whitehorn, P. R., et al. 2012. Neonicotinoid pesticide reduces bumble bee colony growth and queen production. *Science* 336:351–352.

Wilhelm, W. W., et al. 2004. Crop and soil productivity response to corn residue removal: A literature review. *Agronomy Journal* 96:1–17.

7 DEVELOPING SOLUTIONS

Aune, J. B., and A. Coulibaly. 2015. Microdosing of mineral fertilizer and conservation agriculture for sustainable agricultural intensification in sub-Saharan Africa. In *Sustainable Intensification to Advance Food Security and Enhance Climate Resilience in Africa*. Edited by R. Lal et al., Cham, Switzerland: Springer International, pp. 223–234.

Bongaarts, J., and J. Casterline. 2013. Fertility transition: Is sub-Saharan Africa different? *Population and Development Review* 38, Supplement s1, p. 153–168.

Bonsu, M. 1981. Assessment of erosion under different cultural practices on a savanna soil in the northern region of Ghana. In *Soil Conservation: Problems and Prospects*. Edited by R. P. C. Morgan. Chichester, UK: John Wiley & Sons, pp. 247–253.

Buffett, H. G. 2013. *Forty Chances: Finding Hope in a Hungry World*. New York: Simon & Schuster, 443 pp.

Corbeels, M., et al. 2014. *Meta-Analysis of Crop Responses to Conservation Agriculture in Sub-Saharan Africa*. CGIAR Research Program on Climate Change, Agriculture, and Food Security, CCAFS Report No. 12, Copenhagen, 19 pp.

Corbeels, M., et al. 2014. Understanding the impact and adoption of conservation agriculture in Africa: A multi-scale analysis. *Agriculture, Ecosystems & Environment* 187:155–170.

Ekboir, J., K. Boa, and A. A. Dankyi, 2002. *Impact of No-Till Technologies in Ghana*. Economics Program Paper 02-01, International Maize and Wheat Improvement Center (CIMMYT), Mexico, D. F., 32 pp.

Giller, K., et al. 2011. A research agenda to explore the role of conservation agriculture in African smallholder farming systems. *Field Crops Research* 124:468–472.

Giller, K. E., et al. 2009. Conservation agriculture and smallholder farming in Africa: The heretics' view. *Field Crops Research* 114:23–34.

Knowler, D., and B. Bradshaw. 2007. Farmers' adoption of conservation agriculture: A review and synthesis of recent research. *Food Policy* 32:25–48.

Lal, R., et al., eds. 2015. *Sustainable Intensification to Advance Food Security and Enhance Climate Resilience in Africa*. Cham, Switzerland: Springer International, 665 pp.

Marongwe, L. S., et al. 2011. An African success: The case of conservation agriculture in Zimbabwe. *International Journal of Agricultural Sustainability* 9:153–161.

Ngwira, A. R., J. B. Aune, and S. Mkwinda. 2012. On-farm evaluation of yield and economic benefit of short-term maize legume intercropping systems under conservation agriculture in Malawi. *Field Crops Research* 132:149–157.

Ouédraogo, E., A. Mando, and L. Brussaard. 2008. Termites and mulch work together to rehabilitate soils. *LEISA Magazine* 24, no. 2:28.

Pannell, D. J., R. S. Llewellyn, and M. Corbeels. 2014. The farm-level economics of conservation agriculture for resource-poor farmers. *Agriculture, Ecosystems & Environment* 187:52–64.

Rusinamhodzi, L., et al. 2011. A meta-analysis of long-term effects of conservation agriculture on maize grain yield under rain-fed conditions. *Agronomy and Sustainable Development* 31:657–673.

Thierfelder, C., and P. C. Wall. 2009. Effects of conservation agriculture techniques on infiltration and soil water content in Zambia and Zimbabwe. *Soil and Tillage Research* 105:217–227.

Thierfelder, C., and P. C. Wall. 2010. Rotations in conservation agriculture systems of Zambia: Effects on soil quality and water relations. *Experimental Agriculture* 46:309–325.

Thomson, J. A. 2008. The role of biotechnology for agricultural sustainability in Africa. *Philosophical Transactions of the Royal Society B* 363:905–913.

United Nations, Department of Economic and Social Affairs, Population Division. 2013. *Fertility Levels and Trends as Assessed in the 2012 Revision of World Population Prospects.* United Nations Publication ST/ESA/SER.A/349, 20 pp.

8 THE ORGANIC DILEMMA

Ashford, D. L., and D. W. Reeves. 2003. Use of a mechanical roller-crimper as an alternative kill method for cover crops. *American Journal of Alternative Agriculture* 18:37–45.

Bennett, H. H., and W. C. Lowdermilk. 1938. General aspects of the soil-erosion problem. In *Soils & Men, Yearbook of Agriculture 1938.* 75th Congress, 2nd Session, House Document No. 398, United States Department of Agriculture, Washington, D.C.: Government Printing Office, pp. 581–608.

Cavigelli, M. A., et al. 2009. Long-term economic performance of organic and conventional field crops in the mid-Atlantic region. *Renewable Agriculture and Food Systems* 24:102–119.

Davis, A. S. 2010. Cover-crop roller-crimper contributes to weed management in no-till soybean. *Weed Science* 58:300–309.

Davis, A. S., et al. 2012. Increasing cropping system diversity balances productivity, profitability, and environmental health. *PLoS ONE* 7, no. 10: e47149. doi: 10.1371/journal.pone.0047149.

Delate, K., et al. 2003. An economic comparison of organic and conventional grain crops in a long-term agroecological research (LTAR) site in Iowa. *American Journal of Alternative Agriculture* 18:59–69.

Delate, K., et al. 2013. The long-term agroecological research (LTAR) experiment supports organic yields, soil quality, and economic performance in Iowa. *Online, Crop Management*, doi:10.1094/CM-2013-0429-02-RS.

Delate, K., et al. 2015. A review of long-term organic comparison trials in the U.S. *Sustainable Agriculture Research* 4, no. 3:5–14.

Douds, D. D., Jr., et al. 1995. Effect of tillage and farming system upon populations and distribution of vesicular-arbuscular mycorrhizal fungi. *Agriculture, Ecosystems & Enviornment* 52:111–118.

Drinkwater, L. E., P. Wagoner, and M. Sarrantonio. 1998. Legume-based cropping systems have reduced carbon and nitrogen losses. *Nature* 396:262–265.

Green, J. M., and M. D. K. Owen. 2011. Herbicide-resistant crops: Utilities and limitations for herbicide-resistant weed management. *Journal of Agricultural and Food Chemistry* 59:5819–5829.

Leighty, C. E. 1938. Crop rotation. In *Soils & Men, Yearbook of Agriculture 1938*. 75th Congress, 2nd Session, House Document No. 398, United States Department of Agriculture. Washington, D.C.: Government Printing Office, pp. 406–430.

Letter, D. W., R. Seidel, and W. Liebhardt. 2003. The performance of organic and conventional cropping systems in an extreme climate year. *American Journal of Alternative Agriculture* 18:146–154.

Liebhardt, W. C., et al. 1989. Crop production during conversion from conventional to low-input methods. *Agronomy Journal* 81:150–159.

Liebig, M. A., and J. W. Doran. 1999. Impact of organic production practices on soil quality indicators, *Journal of Environmental Quality* 28:1601–1609.

Lockeretz, W., et al. 1978. Field crop production on organic farms in the Midwest. *Journal of Soil and Water Conservation* 33:130–134.

Mäder, P., et al. 2000. Arbuscular mycorrhizae in a long-term field trial comparing low-input (organic, biological) and high-input (conventional) farming systems. *Biology and Fertility of Soils* 31:150–156.

Mäder, P., et al. 2002. Soil fertility and biodiversity in organic farming. *Science* 296:1694–1697.

McGonigle, T. P., M. H. Miller, and D. Young. 1999. Mycorrhizae, crop growth, and crop phosphorus nutrition in maize-soybean rotations given various tillage treatments. *Plant and Soil* 210:33–42.

Mirsky, S. B., et al. 2012. Conservation tillage issues: Cover crop-based organic rotational no-till grain production in the mid-Atlantic region, USA. *Renewable Agriculture and Food Systems* 27:31–40.

Moyer, J. 2011. *Organic No-Till Farming*. Austin, TX: Acres U.S.A., 204 pp.

Moyer, J. 2013. Perspective on Rodale Institute's Farming Systems Trial. *Online. Crop Management* 12, doi:10.1094/CM-2013-0429-03-PS.

Pimentel, D., et al. 2005. Environmental, energetic, and economic comparisons of organic and conventional farming systems. *BioScience* 55:573–582.

Ponisio, L. C., et al. 2015. Diversification practices reduce organic to conventional yield gap. *Proceedings of the Royal Society B* 282: 20141396, doi:10.1098/rspb.2014.1396.

Posner, J. L., J. O. Baldock, and J. L. Hedtcke. 2008. Organic and conventional production systems in the Wisconsin Integrated Cropping Systems Trials: I. Productivity 1990–2002. *Agronomy Journal* 100:253–260.

Reganold, J. 1989. Farming's organic future. *New Scientist* 122:49–52.

Reganold, J. P., L. F. Elliott, and Y. L. Unger. 1987. Long-term effects of organic and conventional farming on soil erosion. *Nature* 330:370–372.

Reganold, J., P., et al. 1993. Soil quality and financial performance of biodynamic and conventional farms in New Zealand. *Science* 260:344–349.

Rillig, M. C. 2004. Arbuscular mycorrhizae, glomalin, and soil aggregation. *Canadian Journal of Soil Science* 84:355–363.

Ryan, G. F. 1970. Resistance of common groundsel to simazine and atrazine. *Weed Science* 18:614–616.

Ryan, M. R., et al. 2009. Weed-crop competition relationships differ between organic and conventional cropping systems. *Weed Research* 49:572–580.

Scow, K. M., et al. 1994. Transition from conventional to low-input agriculture changes soil fertility and biology. *California Agriculture* 48, no. 5:20–26.

Spargo, J. T., et al. 2011. Mineralizable soil nitrogen and labile soil organic matter in diverse long-term cropping systems. *Nutrient Cycling in Agroecosystems* 90:253–266.

Treseder, K. K., and K. M. Turner. 2007. Glomalin in ecosystems, *Soil Science Society of America Journal* 71:1257–1266.

Tu, C., et al. 2006. Responses of soil microbial biomass and N availability to transition strategies from conventional to organic farming systems. *Agriculture, Ecosystems & Environment* 113:206–215.

Tu, C., J. B. Ristaino, and S. Hu. 2006. Soil microbial biomass and activity in organic tomato farming systems: Effects of organic inputs and straw mulching. *Soil Biology & Biochemistry* 38:247–255.

Wright, S. F., and R. L. Anderson. 2000. Aggregate stability and glomalin in alternative crop rotations for the central Great Plains. *Biology and Fertility of Soils* 31:249–253.

Wright, S. F., J. L. Starr, and I. C. Paltineanu. 1999. Changes in aggregate stability and concentration of glomalin during tillage management. *Soil Science Society of America Journal* 63:1825–1829.

Wright, S. F., and A. Upadhyaya. 1996. Extraction of an abundant and unusual protein from soil and comparison with hyphal protein from arbuscular mycorrhizal fungi. *Soil Science* 161:575–586.

Wright, S. F., and A. Upadhyaya. 1998. A survey of soils for aggregate stability and glomalin, a glycoprotein produced by hyphae of arbuscular mycorrhizal fungi. *Plant and Soil* 198:97–107.

9 CARBON COWBOYS

Aase, J. K., and G. M. Schaefer. 1996. Economics of tillage practices and spring wheat and barley crop sequence in northern Great Plains. *Journal of Soil and Water Conservation* 51:167–170.

Baumhardt, R. L., et al. 2009. Cattle grazing effects on yield of dryland wheat and sorghum grown in rotation. *Agronomy Journal* 101:150–158.

Franzluebbers, A. J. 2007. Integrated crop-livestock systems in the southeastern USA. *Agronomy Journal* 99:361–372.

Franzluebbers, A. J., and J. A. Stuedemann. 2008. Early response of soil organic carbon fractions to tillage and integrated crop-livestock production. *Soil Science Society of America Journal* 72:613–625.

Gentry, L. E., et al. 2013. Apparent red clover nitrogen credit to corn: Evaluating cover crop introduction. *Agronomy Journal*, 105:1658–1664.

Herrero, M., et al. 2010. Smart investments in sustainable food production: Revisiting mixed crop-livestock systems. *Science* 327:822–825.

Janzen, H. H. 2001. Soil science on the Canadian prairies—peering into the future from a century ago. *Canadian Journal of Soil Science* 81:489–503.

Liebig, M. A., D. W. Archer, and D. L. Tanaka. 2014. Crop diversity effects on near-surface soil condition under dryland agriculture. *Applied and Environmental Soil Science* 2014, Article ID 703460, doi:10.1155/2014/703460.

Liebig, M. A., D. L. Tanaka, and B. J. Wienhold. 2004. Tillage and cropping effects on soil quality indicators in the northern Great Plains. *Soil & Tillage Research* 78:131–141.

Manson, M. 1899. Observations on the denudation of vegetation—a suggested remedy for California. *Sierra Club Bulletin* 2:295–311.

Montgomery, D. R. 1999. Erosional processes at an abrupt channel head: Implications for channel entrenchment and discontinuous gully formation. In *Incised River Channels*. Edited by S. Darby and A. Simon. Chichester (UK) and New York: John Wiley & Sons, pp. 247–276.

Peterson, G. A., et al. 1998. Reduced tillage and increasing cropping intensity in the Great Plains conserve soil carbon. *Soil Tillage Research* 47:207–218.

Retallack, G. J. 2001. Cenozoic expansion of grasslands and global cooling. *Journal of Geology* 109:407–426.

Retallack, G. J. 2013. Global cooling by grassland soils of the geological past and near future. *Annual Review of Earth and Planetary Sciences* 41:69–86.

Sainju, U. M., et al. 2010. Dryland soil carbon and nitrogen influenced by sheep grazing in the wheat-fallow system. *Agronomy Journal* 102:1553–1561.

Teague, W. R., et al. 2011. Grazing management impacts on vegetation, soil biota and soil chemical, physical and hydrological properties in tall grass prairie. *Agriculture, Ecosystems & Environment* 141:310–322.

Teague, W. R., et al. 2016. The role of ruminants in reducing agriculture's carbon footprint in North America. *Journal of Soil and Water Conservation* 71:156–164.

Van der Heijden, M. G. A., et al. 2006. The mycorrhizal contribution to plant productivity, plant nutrition, and soil structure in experimental grassland. *New Phytologist* 172:739–752.

Van Pelt, R. S., et al. 2013. Field wind tunnel testing of two silt loam soils on the North American Central High Plains. *Journal of Aeolian Research* 10:53–59.

10 INVISIBLE HERDS

Cavaglieri, L., et al. 2005. Biocontrol of *Bacillus subtilis* against *Fusarium verticillioides* in vitro and at the maize root level. *Research in Microbiology* 156:748–754.

Glaser, B. 2007. Prehistorically modified soils of central Amazonia: A model for sustainable agriculture in the twenty-first century. *Philosophical Transactions of the Royal Society B* 362:187–196.

Glaser, B., and J. J. Birk. 2012. State of the scientific knowledge on properties and genesis of Anthropogenic Dark Earths in central Amazonia (*terra preta de Índio*). *Geochimica et Cosmochimica Acta* 82:39–51.

Glaser, B., J. Lehmann, and W. Zech. 2002. Ameliorating physical and chemical properties of highly weathered soils in the tropics with charcoal—a review. *Biology and Fertility of Soils* 35:219–230.

Goyal, S., et al. 1999. Influence of inorganic fertilizers and organic amendments on soil organic matter and soil microbial properties under tropical conditions. *Biology and Fertility of Soils* 29:196–200.

Hammer, E. C., et al. 2015. Biochar increases arbuscular mycorrhizal plant growth enhancement and ameliorates salinity stress. *Applied Soil Ecology* 96:114–121.

Hansen, V., et al. 2015. Gasification biochar as a valuable by-product for carbon sequestration and soil amendment. *Biomass and Bioenergy* 72:300–308.

Laird, D. A. 2008. The charcoal vision: A win-win-win scenario for simultaneously producing bioenergy, permanently sequestering carbon, while improving soil and water quality. *Agronomy Journal* 100:178–181.

Lehmann, J. 2007. Bio-energy in the black. *Frontiers in Ecology and the Environment* 5:381–387.

Lehmann, J., et al. 2003. Soil fertility and production potential. In *Amazonian Dark Earths: Origin, Properties, and Management*. Edited by J. Lehmann, et al. Dordrecht, Netherlands: Springer, pp. 105–124.

Lehmann, J., J. Gaunt, and M. Rondon. 2006. Bio-char sequestration in terrestrial ecosystems—a review. *Mitigation and Adaptation Strategies for Global Change* 11:403–427.

Lehmann, J., et al. 2011. Biochar effects on soil biota—a review. *Soil Biology & Biochemistry* 43:1812–1836.

Liang, B., et al. 2006. Black carbon increases cation exchange capacity in soils. *Soil Science Society of America Journal* 70:1719–1730.

Medina, J. T. 1934. *The Discovery of the Amazon*. American Geographical Society, Special Publication No. 17, New York, 467 pp.

Mukherjee, A., and R. Lal. 2014. The biochar dilemma. *Soil Research* 52:217–230.

Pietikäinen, J., O. Kiikkilä, and H. Fritze. 2000. Charcoal as a habitat for microbes and its effects on the microbial community of the underlying humus. *Oikos* 89:231–242.

Seneviratne, G. 2011. Developed microbial films can restore deteriorated conventional agricultural soils. *Soil Biology & Biochemistry* 43:1059–1062.

Seneviratne, G., and S. A. Kulasooriya. 2013. Reinstating soil microbial diversity in agroecosystems: The need of the hour for sustainability and health. *Agriculture, Ecosystems & Environment* 164:181–182.

Singh, J. S. 2015. Microbes: The chief ecological engineers in reinstating equilibrium in degraded ecosystems. *Agriculture, Ecosystems and Environment* 203:80–82.

Singh, J. S., V. C. Pandey, and D. P. Singh. 2011. Efficient soil microorganisms: A new dimension for sustainable agriculture and environmental development. *Agriculture, Ecosystems & Environment* 140:339–353.

Steinbeiss, S., G. Bleixner, and M. Antonietti. 2009. Effect of biochar amendment on soil carbon balance and soil microbial activity. *Soil Biology & Biochemistry* 41:1301–1310.

Swain, M. R., and R. C. Ray. 2009. Biocontrol and other beneficial activities of *Bacillus subtilis* isolated from cowdung microflora. *Microbiological Research* 164:121–130.

Swarnalakshmi, K., et al. 2013. Evaluating the influence of novel cyanobacterial biofilmed biofertizers on soil fertility and plant nutrition in wheat. *European Journal of Soil Biology* 55:107–116.

Topoliantz, S., J.-F. Ponge, and S. Ballof. 2005. Manioc peel and charcoal: A potential organic amendment for sustainable soil fertility in the tropics. *Biology and Fertility of Soils* 41:15–21.

Wiedner, K., et al. 2015. Anthropogenic dark earth in northern Germany—the Nordic analogue to *terra preta de Índio* in Amazonia. *Catena* 132:114–125.

Woolf, D., et al. 2010. Sustainable biochar to mitigate global climate change. *Nature Communications* 1:56, doi:10.1038/ncomms1053.

11 FARMING CARBON

Albrecht, W. A. 1938. Loss of soil organic matter and its restoration. In *Soils & Men, Yearbook of Agriculture 1938.* 75th Congress, 2nd Session, House Document No. 398, United States Department of Agriculture. Washington, D.C.: Government Printing Office, pp. 347–361.

Baumhardt, R. L., B. A. Stewart, and U. M. Sainju. 2015. North American soil degradation: Processes, practices, and mitigating strategies. *Sustainability* 7:2936–2960.

Chen, G., and R. R. Weil. 2010. Penetration of cover crop roots through compacted soils. *Plant and Soil* 331:31–43.

Cole, C. V. et al. 1997. Global estimates of potential mitigation of greenhouse gas emissions by agriculture. *Nutrient Cycling in Agroecosystems* 49:221–228.

Dean, J. E., and R. R. Weil. 2009. Brassica cover crops for nitrogen retention in the Mid-Atlantic coastal plain. *Journal of Environmental Quality* 38:520–528.

Foley, J. A., et al. 2005. Global consequences of land use. *Science* 309:570–574.

Hudson, B. D. 1994. Soil organic matter and available water capacity. *Journal of Soil and Water Conservation* 49:189–194.

Jobbágy, E. G., and R. B. Jackson. 2000. The vertical distribution of soil organic carbon and its relation to climate and vegetation. *Ecological Applications* 10:423–436.

Köller, K. 2003. Techniques of soil tillage. In A. L. Titi, ed., *Soil Tillage in Agro-ecosystems*. Boca Raton, FL: CRC Press, pp. 1–25.

Lal, R. 1999. Soil management and restoration for carbon sequestration to mitigate the accelerated greenhouse effect. *Progress in Environmental Science* 1:307–326.

Lal, R. 2004. Soil carbon sequestration impacts on global climate change and food security. *Science* 304:1623–1627.

Lal, R. 2006. Enhancing crop yields in the developing countries through restoration of the soil organic carbon pool in agricultural lands. *Land Degradation & Development* 17:197–209.

Lal, R., 2008. Carbon sequestration. *Philosophical Transactions of the Royal Society B* 363:815–830.

Lal, R. 2010. Managing soils and ecosystems for mitigating anthropogenic carbon emissions and advancing global food security. *BioScience* 60:708–721.

Lal, R., et al. 1998. *The Potential of U.S. Cropland to Sequester Carbon and Mitigate the Greenhouse Effect*. Chelsea, MI: Sleeping Bear Press.

Lal, R., et al. 2004. Managing soil carbon. *Science* 304:39.

Lal, R., et al. 2007. Soil carbon sequestration to mitigate climate change and advance food security. *Soil Science* 172:943–956.

Lawley, Y. E., R. R. Weil, and J. R. Teasdale. 2011. Forage radish cover crop suppresses winter annual weeds in fall and before corn planting. *Agronomy Journal* 103:137–144.

Lehmann, J., J. Gaunt, and M. Rondon, 2006. Bio-char sequestration in terrestrial ecosystems—a review. *Mitigation and Adaptation Strategies for Global Change* 11:403–427.

Luo, Z., E. Wang, and O. J. Sun. 2010. Can no-tillage stimulate carbon sequestration in agricultural soils? A meta-analysis of paired experiments. *Agriculture, Ecosystems & Environment* 139:224–231.

Post, W. M., et al. 2004. Enhancement of carbon sequestration in US soils. *BioScience* 54:895–908.

Reicosky, D. C., et al. 2005. Tillage-induced CO_2 loss across an eroded landscape. *Soil and Tillage Research* 81:183–194.

Rodale Institute. 2014. *Regenerative Organic Agriculture and Climate Change*. Kutztown, PA: Rodale Institute, 24 pp.

Ruddiman, W. F. 2005. *Plows, Plagues, and Petroleum: How Humans Took Control of Climate*. Princeton: Princeton University Press, 224 pp.

White, C. M., and R. R. Weil. 2011. Forage radish cover crops increase soil test P surrounding holes created by the radish taproots. *Soil Science Society of America Journal* 75:121–130.

12 CLOSING THE LOOP

Baker, L. A. 2011. Can urban P conservation help to prevent the brown devolution? *Chemosphere* 84:779–784.

Bateman, A., et al. 2011. Closing the phosphorus loop in England: The spatiotemporal balance of phosphorus capture from manure versus crop demand for fertilizer. *Resources, Conservation, and Recycling* 55:1146–1153.

Bitton, G., et al., eds. 1980. *Sludge—Health Risks of Land Application.* Ann Arbor, MI: Ann Arbor Science, 367 pp.

Brown, S., and M. Cotton, M. 2011. Changes in soil properties and carbon content following compost application: Results of on-farm sampling. *Compost Science & Utilization* 19:88–97.

Brown, S., et al. 2011. Quantifying benefits associated with land application of organic residuals in Washington State. *Environmental Science & Technology* 45:7451–7458.

Carter, L. J., et al. 2015. Uptake of pharmaceuticals influences plant development and affects nutrient and hormone homeostases. *Environmental Science & Technology* 49:12,509–12,518.

Cogger, C. G., et al. 2013. Biosolids applications to tall fescue have long-term influence on soil nitrogen, carbon, and phosphorus. *Journal of Environmental Quality* 42:516–522.

Cogger, C. G., et al. 2013. Long-term crop and soil response to biosolids applications in dryland wheat. *Journal of Environmental Quality* 42:1872–1880.

Jenny, H. 1961. *E. W. Hilgard and the Birth of Modern Soil Science.* Pisa: Collana Della Rivista "Agrochmica," 144 pp.

Hargreaves, J. C., M. S. Adl, and P. R. Warman. A review of the use of composted municipal solid waste in agriculture. *Agriculture, Ecosystems & Environment* 123:1–14.

Harrison, E. Z., et al. 2006. Organic chemicals in sewage sludges. *Science of the Total Environment* 367:481–497.

Khaleel, R., K. R. Reddy, and M. R. Overcash. 1981. Changes in soil physical properties due to organic waste applications: A review. *Journal of Environmental Quality* 10:133–141.

King, F. H. 1911. *Farmers of Forty Centuries, or Permanent Agriculture in China, Korea and Japan.* Emmaus, PA: Organic Gardening Press, 379 pp.

Kinney, C. A., et al. 2006. Survey of organic wastewater contaminants in biosolids destined for land application. *Environmental Science & Technology* 40: 7207–7215.

Li, J., and G. K. Evanylo. 2013. The effects of long-term application of organic amendments on soil organic carbon accumulation. *Soil Science Society of America Journal* 77:964–973.

Liebig, J. V. 1863. *The Natural Laws of Husbandry.* New York: Appleton, 387 pp.

MacDonald, G., et al. 2011. Agronomic phosphorus imbalances across the world's croplands. *Proceedings of the National Academy of Sciences* 108:3086–3091.

Metson, G. S., et al. 2016. Feeding the Corn Belt: Opportunities for phosphorus recycling in U.S. agriculture. *Science of the Total Environment* 542:1117–1126.

Mihelcic, J. R., L. M. Fry, and R. Shaw. 2011. Global potential of phosphorus recovery from human urine and feces. *Chemosphere* 84:832–839.

National Research Council (NRC). 2002. *Biosolids Applied to Land: Advancing Standards and Practices.* Committee on Toxicants and Pathogens in Biosolids Applied to Land. Washington, D.C.: National Academies Press, 345 pp.

Paull, J. 2011. The making of an agricultural classic: Farmers of forty centuries or permanent agriculture in China, Korea, and Japan 1911–2011, *Agricultural Sciences* 2:175–180.

Petersen, S. O., et al. 2003. Recycling of sewage sludge and household compost to arable land: Fate and effects of organic contaminants, and impact on soil fertility. *Soil & Tillage Research* 72:139–152.

Roccaro, P., and F. G. A. Vagliasindi. 2014. Risk assessment of the use of biosolids containing emerging organic contaminants in agriculture. *Chemical Engineering Transactions* 37:817–822.

Rogers, H. R. 1996. Sources, behavior and fate of organic contaminants during sewage treatment and in sewage sludges. *The Science of the Total Environment* 185:3–26.

Shenstone, W. A. 1905. *Justus von Liebig: His Life and Work (1803–1873)*, New York: Macmillan, 219 pp.

Song, W. L. 2010. Selected veterinary pharmaceuticals in agricultural water and soil from land application of animal manure. *Journal of Environmental Quality* 39:1211–1217.

Sullivan, D. M., C. G. Cogger, and A. I. Bary. 2007. *Fertilizing with Biosolids.* Pacific Northwest Extension publication 508-E, 15 pp.

Tanner, C. B., and R. W. Simonson. 1993. Franklin Hiram King–pioneer scientist. *Soil Science Society of America Journal* 57:286–292.

Tian, G., et al. 2009. Soil carbon sequestration resulting from long-term application of biosolids for land reclamation. *Journal of Environmental Quality* 38:61–74.

Trlica, A., and S. Brown. 2013. Greenhouse gas emissions and the interrelation of urban and forest sectors in reclaiming one hectare of land in the Pacific Northwest. *Environmental Science & Technology* 47:7250–7259.

Wu, C., et al. 2010. Uptake of pharmaceutical and personal care products by soybean plants from soils applied with biosolids and irrigated with contaminated water. *Environmental Science & Technology* 44:6157–6161.

13 THE FIFTH REVOLUTION

Barański, M., et al. 2014. Higher antioxidant and lower cadmium concentrations and lower incidence of pesticide residues in organically grown crops: A systematic literature review and meta-analyses. *British Journal of Nutrition* 112:794–811.

Brady, M. V. 2015. Valuing supporting soil ecosystem services in agriculture: A natural capital approach. *Agronomy Journal* 107:1809–1821.

Brandes, E., et al. 2016. Subfield profitability analysis reveals an economic case for cropland diversification. *Environmental Research Letters* 11:014009, doi:10.1088/1748-9326/11/1/014009.

Brouder, S. M., and H. Gomez-Macpherson. 2014. The impact of conservation agriculture on smallholder agricultural yields: A scoping review of the evidence. *Agriculture, Ecosystems and Environment* 187:11–32.

Carlisle, L., and A. Miles. 2013. Closing the knowledge gap: How the USDA could tap the potential of biologically diversified farming systems. *Journal of Agriculture, Food Systems, and Community Development* 3:219–225.

Corsi, S., et al. 2012. *Soil Organic Carbon Accumulation and Greenhouse Gas Emission Reductions from Conservation Agriculture: A Literature Review.* Integrated Crop Management vol. 16-2012, Plant Production and Protection Division, Food and Agriculture Organization of the United Nations, Rome, 89 pp.

Crews, T. E., and M. B. Peoples. 2004. Legume versus fertilizer sources of nitrogen: Ecological tradeoffs and human needs. *Agriculture, Ecosystems & Environment* 102:279–297.

De Vries, F. T., et al. 2013. Soil food web properties explain ecosystem services across European land use systems. *Proceedings of the National Academy of Sciences* 110:14,296–14,301.

DeLonge, M. S., A. Miles, and L. Carlisle. 2016. Investing in the transition to sustainable agriculture. *Environmental Science & Policy* 55:266–273.

Federated Farmers of New Zealand. 2005. *Life After Subsidies: The New Zealand Farming Experience—20 Years Later.* Wellington: Federated Farmers of New Zealand, 4 pp.

Fierer, N. 2013. Reconstructing the microbial diversity and function of pre-agricultural tallgrass prairie soils in the United States. *Science* 342:621–624.

Giller, K. E. et al. 2015. Beyond conservation agriculture. *Frontiers in Plant Science* 6:870, doi:10.3389/fpls.2015.00870.

Kassam, A., et al. 2009. The spread of Conservation Agriculture: Justification, sustainability and uptake. *International Journal of Agricultural Sustainability* 7:292–320.

Kibblewhite, M. G., K. Ritz, and M. J. Swift. 2008. Soil health in agricultural systems. *Philosophical Transactions of the Royal Society B* 363:685–701.

Lal, R. 2014. Societal value of soil carbon. *Journal of Soil and Water Conservation* 69:186A–192A.

Lal, R. 2015. A system approach to conservation agriculture. *Journal of Soil and Water Conservation* 70:82A–88A.

Lal, R. 2015. Restoring soil quality to mitigate soil degradation. *Sustainability* 7:5875–5895.

Palm, C. et al. 2014. Conservation agriculture and ecosystem services: An overview. *Agriculture, Ecosytems, and Environment* 187:87–105.

Pretty, J. N., et al. 2006. Resource-conserving agriculture increases yields in developing countries. *Environmental Science & Technology* 40:1114–1119.

Pretty, J. N., J. I. L. Morison, and R. E. Hine. 2003. Reducing food poverty by increasing agricultural sustainability in developing countries. *Agriculture, Ecosystems & Environment* 95:217–234.

Robertson, G. P., et al. 2014. Farming for ecosystem services: An ecological approach to production agriculture. *BioScience* 64:404–415.

Souza, R. C., et al. 2013. Soil metagenomics reveals differences under conventional and no-tillage with crop rotation or succession. *Applied Soil Ecology* 72:49–61.

Syers, J. K. 1997. Managing soils for long-term productivity. *Philosophical Transactions of the Royal Society of London B* 352:1011–1021.

Tsiafouli, M. A., et al. 2015. Intensive agriculture reduces soil biodiversity across Europe. *Global Change Biology* 21:973–985.

Wall, P. C. 2007. Tailoring conservation agriculture to the needs of small farmers in developing countries: An analysis of issues. *Journal of Crop Improvement* 19:137–155.

INDEX

ABOUT THE AUTHOR

David R. Montgomery is a MacArthur Fellow and professor of geo-morphology at the University of Washington. He is an internationally recognized geologist who studies landscape evolution and the effects of geological processes on ecological systems and human societies. An author of award-winning popular-science books, he has been featured in documentary films, network and cable news, and on a wide variety of television and radio programs, including *NOVA*, *PBS NewsHour*, *Fox & Friends*, and *All Things Considered*. When not writing or doing geology, he plays guitar and piano in the band Big Dirt. He lives in Seattle with his wife, Anne Biklé, and their black lab guide-dog drop-out, Loki. Connect with him at www.dig2grow.com or follow him on Twitter (@dig2grow).